Mountain
Flowers
of Britain and Europe

Mountain Flowers
of Britain and Europe

S. STEFENELLI

Translated by Dr Lucia Woodward and
edited by Dr Paul Sterry

David & Charles

ACKNOWLEDGEMENTS

Colour photographs by S. Stefenelli; except plates 108, 154 and 157 by
Jacques de Sloover, and plates 8, 9, 15, 17, 21, 22, 28, 39, 42, 46, 49, 65,
78, 93, 96, 107, 136, 152, 158, 160, 161, 162 and cover photographs by
Paul Sterry, Nature Photographers Ltd.

A DAVID & CHARLES BOOK

First published in Italy by Priuli et Verlucca editori, 1977
Published in English by David & Charles, 1982
Revised English edition published by David & Charles, 1994

A catalogue record for this book is available from the British Library.

ISBN 0 7153 0202 7

Printed in Italy by LEGO SpA, Vicenza
for David & Charles
Brunel House   Newton Abbot   Devon

CONTENTS

INTRODUCTION

It is the sheer beauty of mountain flowers that prompted me to write a book about them; a book that would generate an ever-increasing interest in, and love for, those little plants which can be found in the most varied mountain habitats, from woodland undergrowth to meadow pasture, among stones or in moraines by glaciers, or even within fissures beneath snowy caps.

Each year increasing numbers of walkers and holidaymakers visit the mountains; each year the flowers, however common or rare the species, run the risk of extinction due to indiscriminate and senseless gathering. And senseless it is, because in most cases these flowers—so beautiful in colour and shape—enthusiastically gathered at the beginning of a walk, wither within a few minutes of being picked and are discarded by the path-side.

What is often incredible is the environment in which a small plant manages to grow and produce its splendid, vividly coloured flowers. It often appears among rocks, in a small niche or in a fissure of a rocky face. Despite adverse conditions this tiny plant has managed to develop and survive and we should be encouraged to pause briefly and admire it in its natural environment. And yet, quite often, in the blinking of an eye and with utter irresponsibility, that tiny life is obliterated: the plant will no longer be able to produce again what is often its only flower, and with the flower will vanish its chances of seed and self-perpetuation. Along the path there may be hundreds of flowers similar to the one growing in the rocky face, but these are hardly ever considered: preference is given to the one found in an almost unimaginable spot, and by plucking it one denies others the pleasure of admiring one of nature's miracles. All this because so many people lack all respect for nature, probably because they lack all knowledge of its essence and value.

Hence this book. Many have already been written, but this one is different in that it aims to increase the reader's knowledge of plants and flowers by an easier and more spontaneous system of research rather than by the normal methods of botanical analysis and description. The beginner is guided through the various stages of identification of each plant: from observation of the colour of the flower in question through the use of 73 symbols referring to 17 different peculiarities, together with two scales, to the identification of the plant and its habitat. Thus it becomes easy and entertaining to distinguish the monocotyledons from the dicotyledons; the annuals from the biennials and the perennials; the plant's various forms; its underground organs; the different types and growing habits.

The number of symbols is necessarily limited; consequently, the description of the plant's characteristics as indicated by the symbols is approximate. Identification is made easier by the presence of the common name above each plate. The Latin names are given with their various synonyms; also given are the family names, the colour or colours the plant may assume, and its world-wide distribution. One is also told whether the plant is common or rare, and whether it is endemic, ie restricted to a specific area, itself the result of particular events and ages, and possibly morphologically altered from its original form during the course of time. An endemic plant is not necessarily a rare one; the phenomenon is often widespread within its own habitat or within its distinct geographical areas. Finally, information is given as to whether the plant is poisonous, medicinal or essence-producing, and which parts of the plant are used and what their properties are.

From what has been said so far, one can assume that many of the plants which form the so-called 'alpine flora' are widely distributed, often throughout the world. Can they still be called 'alpines' even though growing on mountains other than the Alps? By 'alpines' we mean all the plants which grow above the treeline whatever the mountain range.

Over the ages, alpine flora has had a very eventful history, particularly as a consequence of glaciations. It has often been excluded from the peaks of its origin and compelled to exist at lower levels, sometimes as far away as the sea, or to 'emigrate' to other areas. As a result of these 'migrations' one can find on the Alps, for instance, plants which originated in distant regions, sometimes from the Mediterranean basin, or as far away as Asia (as in the case of the Edelweiss) and which have had to modify themselves and adapt. Glaciations are also the cause of the formation of many endemic species which have survived within certain microclimates, within oases formed by the movements of the glaciers, by winds, or by general changes of climate. All this, together with the continuing evolution of all living species, causes all vegetal forms to be constantly dynamic. Plants have been known to alter their habitat within living memory: new species can be found where others once lived and have long since gone.

All plants and animals are named and classified using terms which may be incomprehensible and difficult to pronounce. There is a very simple reason for this. In 1735 the greatest naturalist of all time, the well-known Swedish doctor Charles Linnaeus (1707–1778), wrote his *Systema Naturae* (seven volumes), and in it he introduced the concept of family, genus and species, and first used the double nomenclature in Latin which is still in practice today.

His system is of enormous importance to scholars and naturalists throughout the world for, by adopting one official language (Latin) in order to identify and define plants and animals, anyone can understand exactly which species is being discussed.

Following Linnaeus, many other naturalists and botanists used the binomial nomenclature to classify plants according to divisions, classes, orders, families, genera, species, varieties, etc. The binomial genus-species is followed by the name (often abbreviated) of the scientist who established it. This is the reason for the use of the two Latin names, followed by a person's name (or abbreviation), and sometimes by a date indicating the year in which that particular scientist studied the plant and gave it a name to distinguish it permanently from others.

TAXONOMY

Taxonomy (genus and species) is in accordance with Tutin, T. G., Heywood, V. H., Burgess, N. A., Valentine, D. H., Walters, S. M., Webb, D. A., *Flora Europaea*, Vol I (1964), Vol II (1968), Vol III (1972), Vol IV (1976), Cambridge University Press.

Where the plants do not appear in the above four volumes (eg monocotyledons) the taxonomy follows Binz, A. & Thommen, E., *Flore de la Suisse*, 3rd ed. Neuchatel (1966).

Synonyms are based on the following publications:
Fiori, A., *Nuova Flora Analitica d'Italia*, Vols I and II, Bologna (1969). Shown as (**F**).
Binz, A. & Thommen, E., *Flore de la Suisse*, 3rd ed., Neuchatel (1966). Shown as (**BT**).
Hess, H. E., Landolt, E. & Hirzel, R., *Flora der Schweiz und angrenzender Gebiete*, Vols I–III, Basel & Stuttgart (1967–72). Shown as (**HLH**).

Other synonyms are given without quoting the work they have been taken from and are indicated by (*****).

HOW TO READ THE SYMBOLS

Type of plant

1–1. Plant belonging to the Gymnosperms: acicular leaves, often very long, or scale-shaped. Woody stalk, usually with resinous furrows; naked ovules; producing strobiles (cones), galls or berries.

1–2. Dicotyledons: plants producing an embryo with two cotyledons; leaves are net-veined, palm-veined, shield-veined, or veined in other patterns but never with parallel veins. Stems with more complex structure, the vascular system forming annual rings; also present is the cambium, a tissue which allows the diametral growth of the plant and the formation of woody tissues.

1–3. Monocotyledons: having an embryo with only one cotyledon; leaves with parallel veins running lengthwise; stems with relatively simple structure, as the vascular system, lacking the cambium, is haphazardly arranged over the whole of the thickness. Usually herbaceous plants.

Plant cycle

2–1. Annual plant: lasting only one season and completing its life cycle with the production of seeds.

2–2. Biennial plant: surviving for two consecutive years; during the first year it only develops its vegetative system, during the second it produces flowers and fruits with seeds, thus completing its biological cycle.

2–3. Perennial plant: surviving for longer than two years, although not always of a woody nature.

Kind of plant

3–1. Herbaceous: having the consistency of grass.

3–2. Under-shrub: a plant whose stem is woody at the base and herbaceous in the upper parts.

3–3. Shrub or bush: a woody plant, branching out almost from ground level and reaching a height of 1–5m (3–16ft).

11

3–4. Tree: a woody plant of large dimensions, having a columnar trunk with woody ramifications forming the framework of the upper part.

Underground organs

4–A. Root

4–A.1. Tap root: elongated main root from which secondary roots radiate and in turn ramificate; it originates from the collar of the plant directly opposite the stem.

4–A.2. Fasciculate roots: a system of several thin roots, all similar, all originating from the collar.

4–A.3. Fleshy root: a more or less conical root, sometimes spherical, with a more or less distinguishable apex.

4–B. Modified underground stem

4–B.1. Bulb: a shortened underground stem, protected by foliar and fleshy, superimposed organs (cataphylls); fasciculate roots develop from its lower part. Several types exist, among which are the scaly bulbs (*Lilium*), tunicate bulbs (*Allium*), tuberous bulbs (*Crocus*), etc.

4–B.2. Tuber: a shortened and irregular underground stem, thick and globular, fleshy, provided on the outside with buds (eyes) and internally with reserve substances. It is connected with underground branches carrying the roots.

4–B.3. Rhizome: underground stem more or less horizontal, knotty and irregular. It differs from a root in that it is provided with buds from which grow secondary, above-ground stems. The roots grow from the underside of the knots.

Stem growth

5–1. Dwarf plant: stemless and in the shape of a rosette; also with a stem growing to a maximum of 30cm (1ft).

5–2. Cushion-like plant: having tiny, thick-set branches which form a thick conglomerate.

5–3. Spreading plant: having ramifications of various lengths spread out on the soil.

5–4. Stoloniferous plant: a plant with branches spreading on the soil and forming new roots which in turn produce new autonomous plants (including some spreading ones forming occasional roots).

5–5. Erect plant: a plant with one or more stems growing vertically from the roots. Usually taller than 30cm (1ft), bushy or arborescent.

5–6. Rambling plant: with stem(s) initially spreading on the soil before becoming erect.

5–7. Climbing plant: having a generally slender stem which grows by clinging to other plants or to various supports by means of tendrils or discs; also a plant the stem of which grows by twining itself around the support.

Plant sexuality

6–1. Hermaphrodite: a plant with bisexual flowers, ie with male organs (stamens) and female organs (pistils) growing on the same flower.

6–2. Monoecious plant: having male and female flowers growing separately on the same plant.

6–3. Dioecious plant: when the male flowers grow on one individual and the female flowers on another.

Inflorescence
(distribution of the flowers on the plant)

7–1. Single flower: plant with only one flower, or a single flower on each branch or stem.

7–2. Amentum: a generally drooping inflorescence with small sessile flowers densely set on the axis.

7–3. Spike: a usually erect inflorescence carrying sessile flowers along a central axis. The symbol also applies to inflorescences formed by composite spikes.

7–4. Raceme or cluster: inflorescence formed by an elongated stem supporting pedicillate flowers. The symbol includes simple and composite racemes in their various forms: resembling an umbel or a corymb, with ramificated branches, defined or undefined.

7–5. Scorpioid cyme: inflorescence with unilateral development. The symbol covers both simple and composite cymes.

7–6. Corymb: inflorescence formed by flowers with pedicils originating at various heights on the primary axis and all reaching more or less the same level. This symbol includes both simple and composite corymbs, defined or undefined.

7–7. Umbel: inflorescence formed by flowers with pedicils originating from the same point of the main axis and reaching more or less the same height. Both simple and composite inflorescences are covered by the symbol.

7–8. Flat capitulum: the inflorescence is limited to a flat or convex receptacle; the flowers are sessile or have short pedicils, either ligulate or tubulate, often arranged to look like a single flower.

7–9. Oblong capitulum: a globular, oblong or elongated inflorescence, with identical flowers, either sessile or with brief pedicils, closely set along the axis.

Shape of flowers

8–A. Special flowers

8–A.1. Bare flower: flowers having no petals and often no calix, as in many grasses and some trees.

8–A.2. Composite: inflorescence consisting of several flowers either similar (*Hieracium, Crepis*) or dissimilar (*Aster, Arnica*).
Note: this inflorescence is included with the symbols of flowers as beginners might mistake it for a real flower.

8–B. Flowers with dialypetalous corolla
(ie formed by several distinct petals, either regular or irregular)

8–Ba. Regular dialypetalous corolla:

8–Ba.1. Corolla with 4 petals or petaloidal sepals (*Cruciferae*).

8–Ba.2. Corolla with more than 4 petals, tepals or petaloidal sepals (*Rosaceae, Liliaceae, Amaryllidaceae*, etc).

8–Bb. Irregular dialypetalous corolla:

8–Bb.1. Corolla like that of the *Violaceae* (violets, pansies, etc).

8–Bb.2. Papilionaceous (or similar) corolla (*Lathyrus*).

8–C. Flowers with gamopetalous corolla
(The petals are welded at least at the base, thus forming a single organ with more or less marked lobes and regular or irregular corollae.)

8–Ca. Regular gamopetalous corolla:

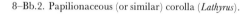

8–Ca.1. Bell-shaped corolla (*Campanula*, etc).

8–Ca2. Corolla with pronounced petaloidal lobes (*Gentianella*, etc).

8–Cb. Irregular gamopetalous corolla:

8–Cb.1. Corolla with bilateral symmetry, ie divided into 2 parts, labiated or 4-lobed: dicotyledons (*Labiatae, Scrophulariaceae*, etc).

8–Cb.2. Corolla with bilateral symmetry, the inferior lobe (labellum) being different from the others: monocotyledons (*Orchidaceae*).

Position of the leaves on the plant

9–1. Rosette: leaves arranged at the base of the plant in variously-shaped rosettes from the centre of which grow one or more stems.

9–2. Alternate: leaves growing on the stem at different levels, never opposed.

9–3. Opposite: leaves growing in pairs on the same level and opposed.

9–4. Whorled: more than two leaves growing at the same level around the stem or branch.

9–5. Imbricated: leaves growing over one another, in well-defined order, like roof tiles, covering one another at the edges.

Shape of leaves

10–1. Narrow leaf: acicular, linear, lanceolate, spatulate.

10–2. Broad leaf: ovate, obovate, ellyptical, round, cordiform, sagittate.

10–3. Single leaf with more or less incised margin or lobes: lobed, pinnatifid, laciniate.

10–4. Single leaf indented almost to the centre in the shape of a fan, or radially: palmate, etc.

10–5. Composite leaf formed by variously-shaped folioles attached to the same axis: trifoliolated, pinnate, bipinnate, etc.

15

Light requirements
(exposure, heliophilia, sciophilia)

  11–1. In full sun.

 11–2. In half shade.

11–3. In full shade.

Humidity requirements (hygrophilia)

 12. Soil humidity: humid or cool soil. If this symbol is not coloured the plant in question is either xerophile or has no particular requirements.

Habitat

 13–1. Meadows, grasslands.

 13–2. Steppes, stony pastures.

 13–3. Grounds covered in rubble or stones.

 13–4. Woodland.

 13–5. Pastures.

 13–6. Peatbogs and fens.

 13–7. Running or still waters.

 13–8. Gravelly shores and banks.

 13–9. Moraines, screes.

 13–10. Rocks and their fissures.

Pedology (characteristics of the soil, pH)

14–1. Basic or alkaline soils, calcareous.

14–2. Neutral soils.

14–3. Acid soils, siliceous, rich in humus and peat.

Phenology (flowering period)

15. Roman figures: indicating the flowering month.

I	II	III	IV	V	VI	VII	VIII	IX	X	XI	XII

Altitude limits

16. Altitudes from 0–4,200m (13,800ft) above sea level: the minimum and maximum levels at which a plant can normally be found.

Plant height

17. Scales from 0–120cm (4ft) and from 0–12m (40ft): indicating the minimum and maximum height a given plant can reach. Heights above 12m (40ft; trees) are indicated separately in the text.

Plant properties

Poisonous plant

Medicinal plant

Essence-producing plant

HABITAT

Climatic conditions in mountainous areas vary greatly from the plains to the snow-covered summits, providing a wide variety of habitats and thus very many different species of mountain flowers. The rambler or climber will know how changeable the scenery can be in the course of a walk or climb. At the outset one encounters leafy forests which are shady and cool; then come the conifers—first firs, then spruces, pines and larches—replacing the beeches to form a tall, dark and heavily-scented forest; higher up, the forest opens up and the trees become sparse and stunted. Once the last sparse trees have been left behind, one encounters scrubland containing shrubs and rhododendrons. As the summit is approached, the grass cover becomes patchy and shorter; stretches of grass alternate with patches of stone, moraines alternate with the first patches of ice, followed finally by bare rock or glacier.

Together with its climate and flowers, each type of scenery characterises what is called a *vegetation zone*. Following the recognised classification system, the area of plains and hills at the foot of the mountains is the *colline zone* which occurs up to 550–1,200m (1,800–4,000ft). This is the domain of food crops such as corn and maize, of vines and of oak forests, and extends to the limit of vine culture. Above this is the *montane zone* with its meadows and beech groves which extends to the limit of deciduous woods at 800–1,700m (2,600–5,600ft). The *subalpine zone* is the almost exclusive domain of the conifer—firs, pines and larches—up to the limit of coniferous woods at 1,600–2,400m (5,200–7,900ft). The region of the mountain pine, followed by dwarf shrubs and alpine meadows up to 1,700–2,800m (5,600–9,200ft), is the *alpine zone*. The shrubs include Alders, and there are under-shrubs such as Bearberry and Whortleberry. Finally, as the meadows become sparser and sparser, the *snow zone* is reached, above the climatic snow line, where in summer only a few lichens and mosses grow in the areas uncovered by snow. Against this background mountain flowers take their place.

Plate 1 The resinous forest covers the steep slopes of valleys between 1,700 and 2,200–2,400m (5,600 and 7,200–7,900ft). Some mountain flowers, such as the Twinflower (pl 16), the One-flowered wintergreen (pl 14) or the Large speedwell (pl 86), will only grow in the shelter of this forest. Clearings interrupt the forest to make way for the meadows and pastures (towards the bottom of the plate). It is in these stretches of grass that the Globeflower (pl 32) flourishes, as well as Arnica (pl 53),

Purple gentian (pl 120), Mountain cornflower (pl 151) and many others. Higher up, the more rugged climatic conditions prevent trees from growing and the forest grows thinner and finally disappears. The natural boundary of the trees appears towards the top of the plate; it varies in altitude according to the exposure and the latitude of the mountain range.

Plate 2   At the foot of the slope of this valley the coniferous forest dies away on the steep banks of the rushing stream; here can be found Masterwort (pl 13), White butterbur (pl 24) and Alpine butterbur (pl 87). Beyond the stream the grass, which is scattered with stunted conifers, is nothing like that of the meadow—it is a peat-bog, one of the most fascinating environments both from the botanical and the aesthetic point of view. The origins of this habitat, which is less widespread than the others, are to be found in the slow accumulations over the centuries of the residues of bog plants which still continue to pile up and thus raise the level of the peat bog; numerous species can be found growing there, such as Grass-of-Parnassus (pl 11), Common cottongrass (pl 27), Bird's-eye primrose (pl 83) and Early marsh orchid (pl 129), mixed with mosses or—where the soil is acid—peat-moss.

Plate 3   A high alpine valley, like that which leads to the Lauson pass (3,296m, 10,800ft), provides yet further habitats for mountain flowers. At the bottom, at a height of about 1,700m (5,600ft), a grazing meadow is cut out of the coniferous forest.

The forest climbs the first broken buttresses of the rocky outcrops but, higher up, the scattered trees give way to the stony alpine meadows, home of the Alpine anemone (pl 4), Vitaliana (pl 51), Alpine catchfly (pl 62), Least Primrose (pl 66), Black vanilla orchid (pl 130) and Trumpet gentian (pl 131). Then come the snow slopes which, with the moraines, climb to the *névé*. On the right, some black rocky escarpments jut out; their crevices may shelter Rough hawkweed (pl 8), Yellow wood-violet (pl 48) and Black speedwell (pl 146). The great white scar which mutilates the scenery to the right of the rushing stream is a former quarry of *lauzes*—thin slabs of rock used to cover roofs. Vegetation has not yet succeeded in recolonising the quarry although it has been disused for over twenty years.

Plate 4   Scree, rocky outcrops and short grass are the habitat of the most typical and attractive alpine flowers—delicate plants, in cushions or little tufts, like Alpine rock jasmine (pl 84), Purple saxifrage (pl 113), and Mt Cenis bellflower (pl 149). In the little valleys and depressions in high-altitude habitats (see centre of plate) the snow takes longest to thaw, and these areas contain a highly specialised vegetation with more lichens and mosses than plants with flowers. These plants of the snow zone, such as Sibbaldia (pl 44), Retuse-leaved saxifrage (pl 119) and Least willow (pl 156), have only a few weeks to flower between the thaw which comes late at this altitude and the autumn which arrives particularly early. Although much reduced in height, these plants are still perennials, and may live for many years in constant development.

Therefore each of their summers, very short and precarious, is used to produce leaves and flowers which have been developing since the preceding autumn and to prepare buds which will lie dormant all winter, sheltered from the heavy frosts by the thick mantle of snow.

THE FIELD GUIDE

The species have been grouped together according to their dominant colour. For ease of reference, these colours are indicated at the edge of the page.

Cerastium alpinum L.
Cerastium alpinum L. var. *uniflorum* (Thom.) Fiori (**F**)
Cerastium uniflorum Clairv. (**BT**) (**HLH**)

Family: Caryophyllaceae
Colour: white
Distribution: Alps, Carpathians, N Wales, N England, Scotland
Frequency: locally common

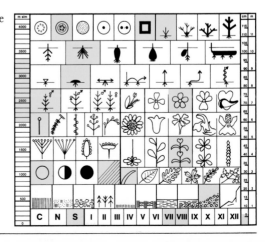

Actaea spicata L., 1753
Actaea spicata L. (**F**) (**BT**) (**HLH**)

Family: Ranunculaceae
Colour: white
Distribution: Alps, Apennines; in Britain, confined to upland limestone areas from Yorkshire to Cumbria; temperate, cold and mountain areas of the northern hemisphere
Frequency: uncommon
Properties: (*roots*) emetic, purgative, sudorific, anti-bronchitis; *external use:* odontalgic

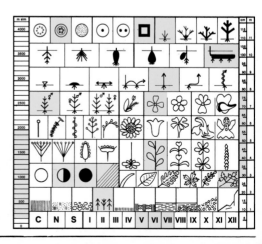

Anemone baldensis L., 1767

Anemone baldensis Turra in L. (**F**)
Anemone baldensis Turra (**BT**) (**HLH**)

Family: Ranunculaceae

Colour: white, pinkish white

Distribution: Alps, Pyrenees, Croatia, Carpathians

Frequency: uncommon

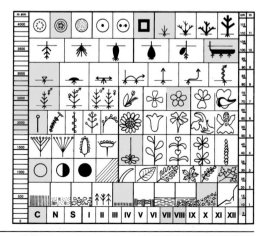

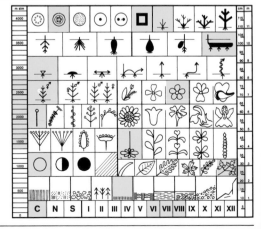

Pulsatilla alpina (L.) Delarbre, 1800, subsp. ***alpina***
Anemone alpina L. var. *typica* (**F**)
Pulsatilla alpina (L.) Delarbre (**BT**)
Pulsatilla alpina (L.) Schrank (**HLH**)

Family: Ranunculaceáe

Colour: white
Distribution: Alps,
Central Apennines,
Pyrenees, Balkans,
Carpathians, Caucasus
Frequency: common
Properties: (*flowers and leaves*) antispasmodic,
sedative

Ranunculus glacialis L., 1753

Ranunculus glacialis L. var. *typicus* (**F**)
Ranunculus glacialis L. (**BT**) (**HLH**)
Oxygraphis vulgaris Freyn (*****)

Family: Ranunculaceae

Colour: white, pink, purple, reddish brown
Distribution: Alps, Sierra Nevada, Pyrenees, Scandinavia, Iceland, Spitzbergen, Carpathians, Alaska, Kola peninsula
Frequency: common
Properties: (*whole flowering plant*) febrifuge

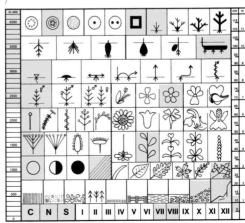

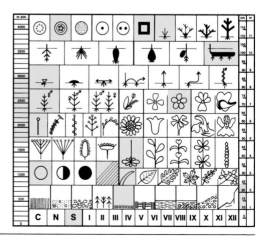

Ranunculus pyrenaeus L., 1771, subsp. ***pyrenaeus***

Ranunculus pyrenaeus L. var. *typicus* (**F**)
Ranunculus pyrenaeus L. (**BT**) (**HLH**)

Family: Ranunculaceae

Colour: white
Distribution: Spain, Pyrenees, Alps, Corsica
Frequency: uncommon

Thlaspi alpestre L.

Thlaspi montanum var. *alpinum* (Crantz) Fiori (**F**)
Thlaspi alpinum Crantz (**BT**) (**HLH**)

Family: Cruciferae

Colour: white, yellowish white
Distribution: Alps, Tuscan Apennines; in Britain, confined mainly to limestone areas in Pennines
Frequency: uncommon

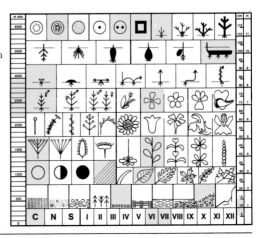

Saxifraga stellaris L.
Saxifraga bryoides L. var. *aspera* (L.) Fiori (**F**)
Saxifraga aspera L. (**BT**) (**HLH**)

Family: Saxifragaceae

Colour: white
Distribution: mountains of central and northern Europe; in Britain, found in upland regions from Snowdonia and the Pennines north to Scotland
Frequency: locally common

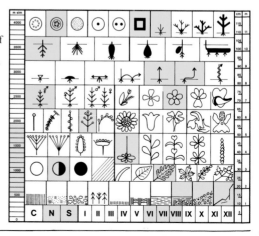

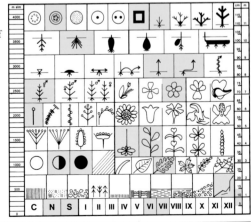

Saxifraga cernua L.
Saxifraga granulata L. var. *typica* (**F**)
Saxifraga granulata L. (**BT**) (**HLH**)

Family: Saxifragaceae

Colour: white
Distribution: mountains of central and northern Europe; in Britain, confined to Scottish highlands
Frequency: very local

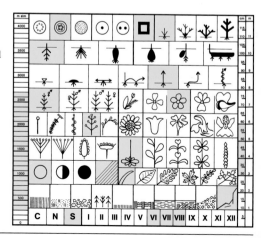

Saxifraga cotyledon L., 1753
Saxifraga cotyledon L. (**F**) (**BT**) (**HLH**)

Family: Saxifragaceae

Colour: white
Distribution:
Piedmontese, Lombard and
Tyrolean Alps, Aosta
Valley, Pyrenees, Norway,
Lapland, Iceland,
Carpathians
Frequency: uncommon
(endemic)

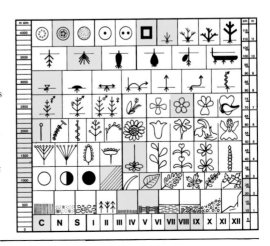

Parnassia palustris L., 1753, subsp. ***palustris***
Parnassia palustris L. (**F**) (**BT**) (**HLH**)

Family: Parnassiaceae

Colour: white
Distribution: Alps, northern and central Apennines and northern Europe; in Britain, found mainly in northern regions and absent from most southern districts
Frequency: locally common
Properties: (*whole plant*) astringent, antidiarrhoeic, antihaemorrhagic, diuretic

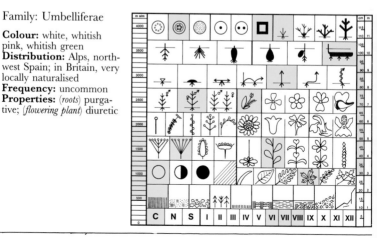

Astrantia major L., 1753, subsp. ***carinthiaca*** Arcangeli, 1882
Astrantia major L., var. *nigra* (Scop.) Fiori (**F**)
Astrantia major L. (**BT**) (**HLH**)

Family: Umbelliferae

Colour: white, whitish pink, whitish green
Distribution: Alps, north-west Spain; in Britain, very locally naturalised
Frequency: uncommon
Properties: (*roots*) purgative; (*flowering plant*) diuretic

Peucedanum ostruthium (L.) Koch, 1824

Peucedanum ostruthium Koch var. *typica* (**F**)
Peucedanum ostruthium (L.) Koch (**BT**) (**HLH**)
Imperatoria ostruthium L. (*****)

Family: Umbelliferae

Colour: white, pink
Distribution: Alps, north-west Spain; in Britain, mainly central and northern counties
Frequency: uncommon
Properties: (*roots*) bitter-tonic, carminative, diaphoretic, expectorant (bronchial system); *external use:* chewed to relieve toothache and migraine; liqueur manufacture

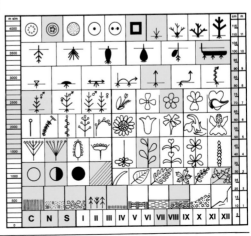

Moneses uniflora (L.) A. Gray, 1848
Pirola uniflora L. (**F**)
Pyrola uniflora L. (**BT**) (**HLH**)

Family: Pyrolaceae

Colour: white
Distribution: Alps, central
and northern Europe,
Caucasus, North America;
in Britain, confined to
Scottish highlands
Frequency: rare
Properties: (*leaves*)
astringent, diuretic

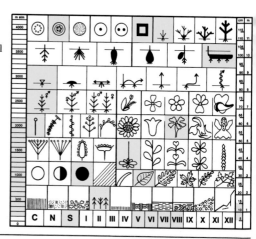

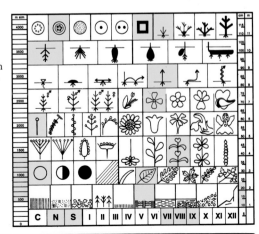

Cornus suecica L.
Chamaepericlymenum suecicum (L.)

Family: Cornaceae

Colour: white

Distribution: northern Europe and mountain ranges in central Europe; in Britain, found mainly from northern England to Scotland on upland moors

Frequency: locally common

Linnaea borealis L., 1753
Linnaea borealis L. (**F**) (**BT**) (**HLH**)

Family: Caprifoliaceae
Colour: white, pinkish white, pink
Distribution: Alps, central and northern Europe, Caucasus, northern Asia, North America; in Britain, confined to Scottish highlands
Frequency: locally common

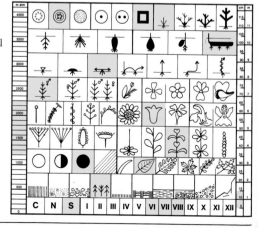

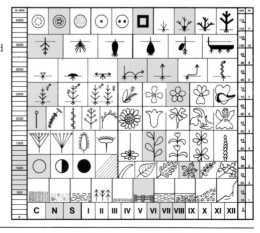

Rubus chamaemorus L.

Family: Rosaceae

Colour: white
Distribution: northern
Europe; in Britain, confined
to north Wales, northern
England and Scotland
Frequency: locally
common

Aster bellidiastrum (L.) Scop. 1769
Bellidiastrum michelii Cass. var. *typicum* (**F**)

Family: Compositae

Colour: white, pinkish
white, purplish white
Distribution: Alps, Apuan
Alps, northern and central
Apennines, Carpathians,
Balkans
Frequency: common

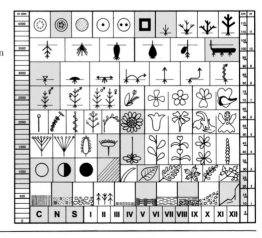

Erigeron uniflorus* L., 1753*
*Erigeron alpinus L. var. uniflorus (L.) Fiori (**F**)*
*Erigeron uniflorus L. (**BT**) (**HLH**)*

Family: Compositae

Colour: white, lilac, pink
Distribution: Alps, Apuan
Alps, central Apennines,
arctic Europe, Pyrenees,
Balkans, Carpathians,
Caucasus, Tibet, Urals,
arctic Asia, Greenland,
North America (Arctic)
Frequency: not common
Properties: (*flower heads*)
antidiarrhoeic, anti-gout,
antirheumatic, balsamic,
diuretic

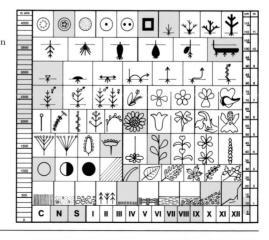

Leontopodium alpinum Cass., 1822

Leontopodium alpinum Cass. var. *typicum* (**F**)
Leontopodium alpinum Cass. (**BT**) (**HLH**)

Family: Compositae

Colour: white
Distribution: Alps, northern Apennines, Pyrenees, Carpathians, Balkans, Croatia, Himalayas, Mongolia, Altais
Frequency: uncommon
Properties: (*leaves and flowers*) astringent, bechic

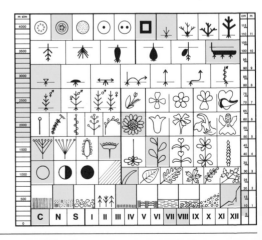

Polygonum viviparum L.

Family: Polygonaceae
Colour: white
Distribution: northern
Europe and upland regions
of central and southern
Europe; in Britain, from
north Wales and northern
England northwards
Frequency: locally
common

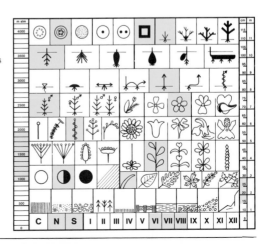

Arenaria norvegica Gunn

Family: Caryophyllaceae

Colour: white

Distribution: northern Europe including Scandinavia and Iceland; in Britain, confined to Scottish highlands and islands, and limestone areas in northern England

Frequency: rare

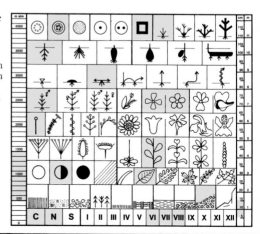

Chrysanthemum alpinum L.

Chrysanthemum alpinum L. (**F**) (**BT**) (**HLH**)
Pyretrum alpinum (L.) Schrank (*)
Tanacetum alpinum (L,) Schultz Bip. (*)

Family: Compositae

Colour: white, pinkish white
Distribution: Alps, Pyrenees, Carpathians
Frequency: not common (endemic)

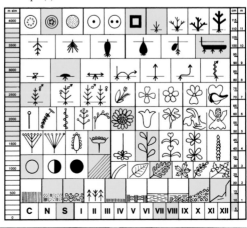

Petasites albus (L.) Gaertner, 1791
Petasites albus Gaertn. (**F**)
Petasites albus (L.) Gaertner (**BT**) (**HLH**)

Family: Compositae

Colour: white, yellowish white

Distribution: Alps, Apennines, Europe, central Russia, Balkans, Caucasus, Armenia, Altais

Frequency: common

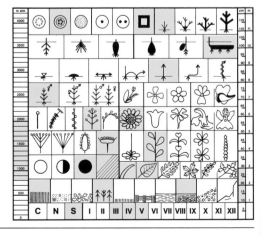

Carlina acaulis L., 1753, subsp. ***acaulis***
Carlina acaulis L. var. *typica* (**F**)
Carlina acaulis L. (**BT**) (**HLH**)

Family: Compositae

Colour: white, purple, brownish purple
Distribution: Alps, Apennines, Europe (from Spain to Russia)
Frequency: common (endemic)
Properties: (*roots*) bitter stomachic, carminative, diaphoretic, antipyretic, diuretic; to be used with caution as it is an irritant

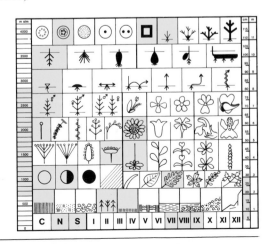

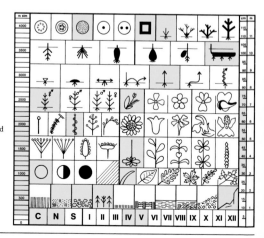

Stipa pennata L., 1753
Stipa pennata L. var. *mediterranea* Trin. & Rupr. (**F**)
Stipa gallica (Steven) Celak. (**HLH**)

Family: Graminaceae

Colour: white, yellowish white, silvery white
Distribution: Mediterranean basin, central Europe, western Asia, Siberia, Africa (Algeria)
Frequency: rare

Note: the feathery awns cause the seed to be transported by the wind in July–August

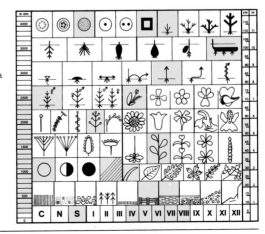

Eriophorum angustifolium Honckeny

Eriophorum polystachyum L., 1753, var. *angustifolium* (Roth, 1789) Fiori (**F**)
Eriophorum angustifolium Honckeny (**HLH**)

Family: Cyperaceae

Colour: white
Distribution: Apennines, Europe, Britain, Siberia, Caucasus, North America, Greenland, southern Africa
Frequency: common

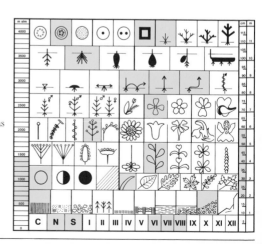

Galium boreale L.

Family: Rubiaceae

Colour: white
Distribution: northern and eastern Europe and mountains in the south; in Britain, it is confined to northern England and Scotland
Frequency: uncommon
Property: formerly used as a bedding material when dried

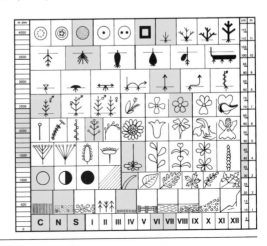

Paradisia liliastrum (L.) Bertol. 1839
Paradisia liliastrum Bert. (**F**)
Paradisia liliastrum (L.) Bert. (**HLH**)

Family: Liliaceae
Colour: white
Distribution: Alps, northern and central Apennines, Pyrenees
Frequency: rare

Allium ursinum L., 1753
Allium ursinum L. (**F**) (**HLH**)

Family: Liliaceae

Colour: white
Distribution: Britain,
Europe, Asia Minor,
Caucasus, Siberia, Ukraine,
Kamchatka (Russia)
Frequency: very common
Properties: (*whole plant*)
depurative, gastric
stimulant, antihelminthic,
hypotensive, diuretic,
antiseptic, antiputrefying,
rubefacient

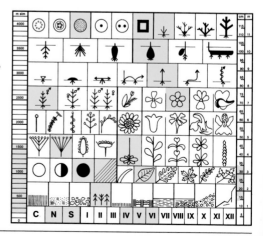

Platanthera bifolia (L.) L. C. M. Richard, 1817

Platanthera bifolia Rich. var. *typica* (**F**)
Platanthera bifolia (L.) Rich. (**HLH**)
Orchis bifolia L. (*****)

Family: Orchidaceae

Colour: white, greenish white
Distribution: Britain, Europe, Sicily, Sardinia, Asia Minor, Caucasus, Himalayas, north-west Africa
Frequency: uncommon

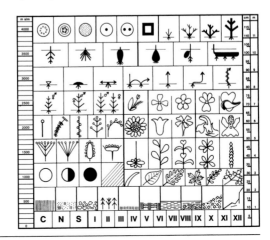

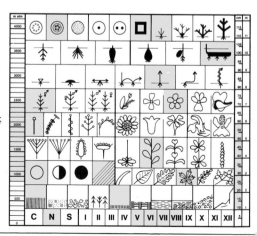

Trollius europaeus L., 1753
Trollius europaeus L. var. *typicus* (**F**)
Trollius europaeus L. (**BT**) (**HLH**)

Family: Ranunculaceae

Colour: yellow
Distribution: Alps,
northern and central
Apennines, Britain, Europe,
Carpathians, Caucasus,
arctic coasts of Russia
Frequency: locally
common
Properties: (*root*) purgative;
slightly poisonous

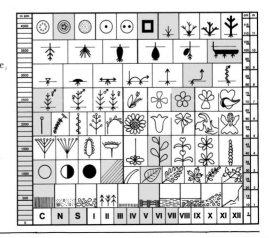

Caltha palustris L., 1753
Caltha palustris L. (**F**) (**BT**) (**HLH**)

Family: Ranunculaceae

Colour: deep yellow, golden yellow
Distribution: Alps, Apennines, Britain, Europe, as far as the Arctic, Caucasus, Himalayas, Siberia, Japan, North America
Frequency: locally common

Aconitum vulparia Reichenb., 1819
Aconitum lycoctonum L. (**F**) (**BT**)
Aconitum vulparia Rchb. (**HLH**)

Family: Ranunculaceae

Colour: yellow, greenish yellow
Distribution: Alps, Apennines, Europe, central Asia as far as Japan, northern Africa (Atlas)
Frequency: uncommon

Note: very poisonous

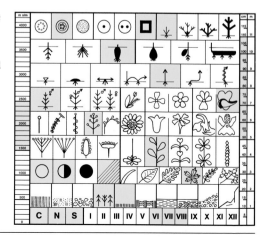

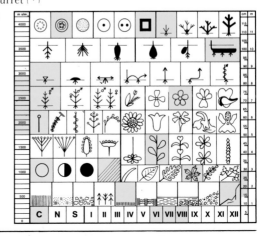

Ranunculus montanus Willd. 1800

Ranunculus montanus W. (**F**)
Ranunculus montanus Willd. (**BT**) (**HLH**)
Ranunculus geranifolius Pourret (*)

Family: Ranunculaceae

Colour: yellow
Distribution: Alps, northern and central Apennines, Jura, Black Forest
Frequency: common

Erysimum helveticum (Jacq.) DC. in Lam. & DC., 1805
(incl. ***Erysimum pumilum*** auct.)
Erysimum hieracifolium L. var. *pumilum* (DC.) Fiori (**F**)
Erysimum pumilum Gaudin (**HLH**)

Family: Cruciferae
Colour: yellow
Distribution: Maritime
Alps, Carinthia, Pyrenees,
Balkans
Frequency: rare

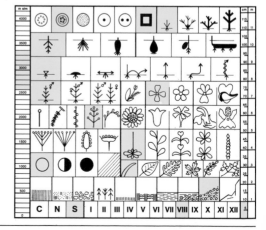

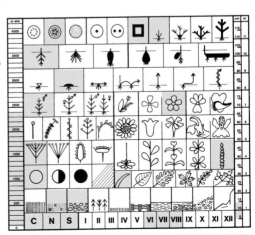

Sedum acre L., 1753
Sedum sexangulare L. var. *acre* (L.) Fiori (**F**)
Sedum acre L. (**BT**) (**HLH**)

Family: Crassulaceae

Colour: yellow
Distribution: Apennines, Sicily, Elba, Britain, Europe, northern Asia, Asia Minor, northern Africa, North America.
Frequency: common
Properties: (*aerial parts*) diuretic, hypotensive, anti-arteriosclerotic

Saxifraga bryoides L., 1753

Saxifraga bryoides L. var. *typica* (**F**)
Saxifraga aspera L. ssp. *bryoides* (L.) Gaudin (**F**)
Saxifraga bryoides L. (**HLH**)

Family: Saxifragaceae

Colour: pale yellow, whitish
Distribution: Alps, Pyrenees, Carpathians, Balkans
Frequency: uncommon

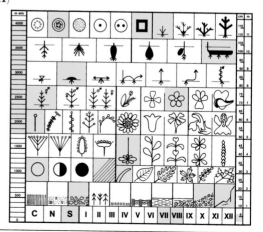

Saxifraga aizodes L.

Family: Saxifragaceae

Colour: yellow
Distribution: northern
Europe and mountains in
central Europe; in Britain,
confined to northern
England and Scotland
Frequency: locally
common

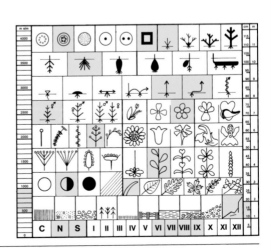

Geum reptans L., 1753

Geum reptans L., (**F**) (**HLH**)
Sieversia reptans (L.) R. Br. (**BT**)
Parageum reptans (L.) Kral (*)

Family: Rosaceae

Colour: yellow
Distribution: Alps, Carpathians, Balkans
Frequency: rare
Properties: (*leaves*) astringent, vulnerary

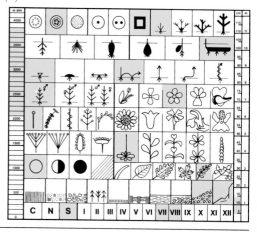

Geum montanum L., 1753

Geum montanum L. (**F**) (**HLH**)
Sieversia montana (L.) R. Br. (**BT**)
Parageum montanum (L.) Hara (*)

Family: Rosaceae

Colour: yellow
Distribution: Alps, Apuan Alps, Tuscan Apennines, Pyrenees, Balkans, Carpathians, Corsica
Frequency: common
Properties: (*roots*) bitter tonic, digestive, astringent

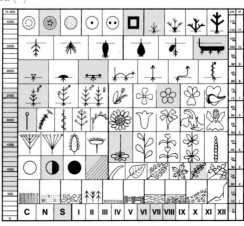

Potentilla fruticosa L.
Potentilla anserina L. (**F**) (**BT**)
Potentilla anserina L. (**HLH**)

Family: Rosaceae

Colour: yellow
Distribution: mountains in
northern and central
Europe; in Britain, confined
to Teesdale and Cumbria
Frequency: very local
Properties: (*rhizome*)
antidiarrhoeic, astringent,
haemostatic, gastric tonic

Potentilla crantzii (Crantz) G. Beck ex Fritsch, 1897

Potentilla verna L. var. *Crantzii* (G. Beck) Fiori (**F**)
Potentilla Crantzii (Crantz) Beck (**BT**) (**HLH**)
Potentilla villosa Zimmeter (*)
Potentilla alpestris Haller fil. (*)
Potentilla salisburgensis Haenke (*)

Family: Rosaceae

Colour: yellow
Distribution: Alps, north and central Apennines, Britain (up to 1,100m [3,600ft]), Europe, western and northern Asia, North America, Greenland
Frequency: locally common

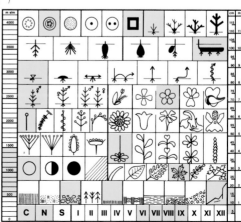

Sibbaldia procumbens L., 1753
Potentilla sibbaldi Hall. f. (**F**)
Sibbaldia procumbens L. (**BT**) (**HLH**)

Family: Rosaceae

Colour: yellow, greenish yellow, pale yellow
Distribution: Alps, central Apennines, Britain (to 1,300m), Europe, central and northern Asia, North America
Frequency: uncommon

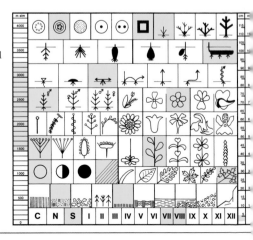

Lotus corniculatus L., 1753
Lotus corniculatus L. var. *arvensis* (Pers) Fiori (**F**)
Lotus corniculatus L. (**BT**) (**HLH**)

Family: Leguminosae

Colour: yellow
Distribution: throughout Europe, central and northern Asia, north Africa (Ethiopia), Japan, Australia
Frequency: very common
Properties: (*flowers*) tranquillizer of the nervous, psychic and cardiac systems, antispasmodic, anticolitic

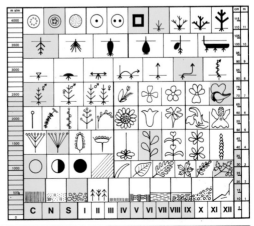

Oxytropis campestris L.

Family: Leguminosae

Colour: pale yellow
Distribution: northern
Europe and mountains in
central Europe; in Britain,
confined to Scottish
mountains
Frequency: very local

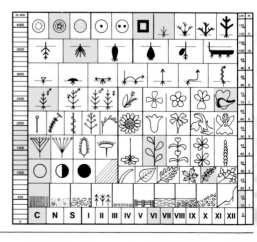

Hippocrepis comosa L., 1753

Hippocrepis comosa L. var. *typica* (**F**)
Hippocrepis comosa L. (**BT**) (**HLH**)

Family: Leguminosae

Colour: yellow
Distribution: southern and central Europe north to central England
Frequency: locally common

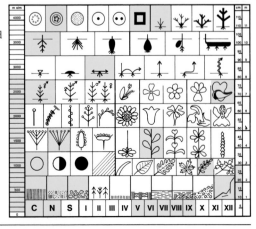

Viola biflora L., 1753
Viola biflora L. (**F**) (**BT**) (**HLH**)

Family: Violaceae

Colour: yellow
Distribution: Alps, Apuan Alps, northern Apennines, Europe, Caucasus, northern Asia, Himalayas, China, Japan, North America
Frequency: common
Properties: (*roots*) emetic, purgative; (*leaves and flowers*) depurative (blood), antiscrofulous, diaphoretic, bechic, expectorant, diuretic, laxative, emollient

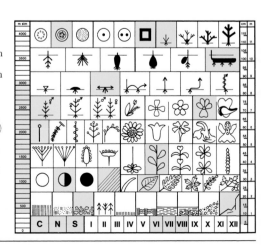

Lilium pyrenaicum Gouan

Family: Liliaceae

Colour: yellow
Distribution: mountains in southern Europe; introduced into parts of southern England
Frequency: local

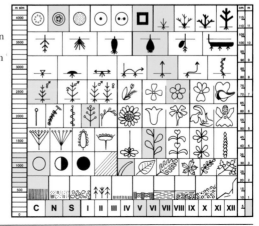

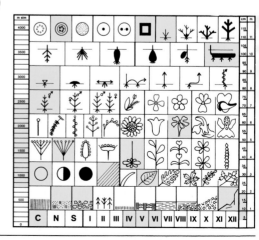

Primula auricula L., 1753

Primula auricula L. var. *Balbisii* (Lehm.) Fiori (**F**)
Primula auricula L. (**BT**) (**HLH**)
Primula ciliata Moretti (*****)

Family: Primulaceae

Colour: yellow, sulphur yellow
Distribution: Alps, northern and central Apennines, Carpathians
Frequency: local

Vitaliana primuliflora Bertol. 1835
subsp. ***canescens*** O. Schwarz, 1963

Douglasia vitaliana Pax (**F**)
Douglasia vitaliana (L.) Pax (**BT**)
Androsace vitaliana (L.)
Lap. (**HLH**)
Gregoria vitaliana (L.)
Duby (*)
Aretia vitaliana Murray
(*)

Family: Primulaceae

Colour: yellow
Distribution: Alps, central
Apennines (Abruzzo),
Pyrenees
Frequency: rare

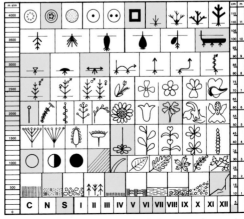

Gentiana lutea L., 1753
Gentiana lutea L. (**F**) (**BT**) (**HLH**)

Family: Gentianaceae

Colour: yellow
Distribution: Alps,
Apennines, Sardinia, Spain,
Balkans, Carpathians
Frequency: common
Properties: (*roots*)
antifermentative, aperitive,
bitter digestive tonic,
febrifuge, vermifuge

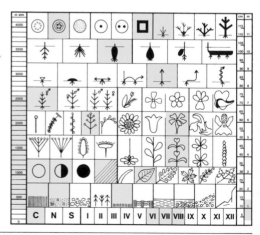

Arnica montana L., 1753
Arnica montana L. var. *typica* (**F**)
Arnica montana L. (**BT**) (**HLH**)

Family: Compositae

Colour: yellow, orangey yellow
Distribution: Alps, northern Apennines, Britain, Europe, Carpathians
Frequency: common
Properties: (*rhizome*) rubefacient: (*flowers*) antiseptic, tonic, digestive, stimulant of the nervous and cardiovascular systems, hypertensive, haemosolvent, febrifuge, sudorific; must be used with great caution as it is poisonous

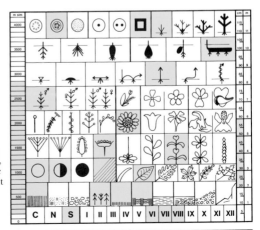

Doronicum grandiflorum Lam., 1786

Doronicum scorpioides Lam. var. *grandiflorum* (Lam.) Fiori (**F**)
Doronicum grandiflorum Lam. (**BT**) (**HLH**)
Arnica scorpioides Jacq. (*)

Family: Compositae

Colour: yellow
Distribution: Alps,
Pyrenees, Balkans
Frequency: uncommon

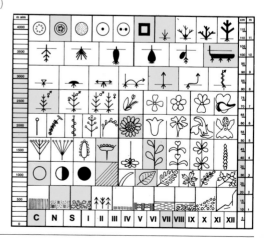

Senecio doronicum (L.) L., 1759
Senecio doronicum L. var. *glabratus* Heg. & Heere (**F**)
Senecio doronicum L. (**BT**) (**HLH**)

Family: Compositae

Colour: yellow, golden yellow, orangey yellow
Distribution: Alps, northern and central Apennines, central and southern Europe, Balkans, Carpathians
Frequency: common

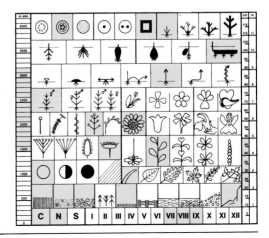

Cirsium spinosissimum (L.) Scop., 1769

Cirsium spinosissimum Scop. var. *typicum* (**F**)
Cirsium spinosissimum (L.) Scop. (**BT**) (**HLH**)

Family: Compositae

Colour: yellow, whitish yellow
Distribution: Alps, Carpathians
Frequency: common

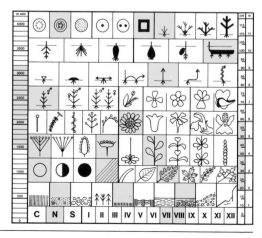

Tolpis staticifolia (All.) Schultz Bip., 1861
Hieracium staticifolium All. (**F**) (**BT**) (**HLH**)

Family: Compositae
Colour: yellow, sulphur yellow
Distribution: Alps, Jura, Balkans, eastern Carpathians, Albania
Frequency: common

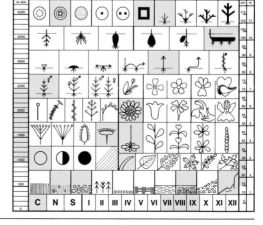

Tragopogon pratensis L., 1753
Tragopogon pratensis L. (**F**) (**BT**) (**HLH**)

Family: Compositae

Colour: yellow, pale yellow
Distribution: Britain, central and southern Europe, southern Russia, Caucasus, western Iran, Siberia, North America (naturalised)
Frequency: uncommon
Properties: (*roots*) edible, aperitive, sudorific, depurative, astringent

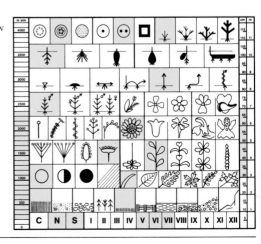

Hieracium peleterianum Mérat, 1812
Hieracium pilosella L. var. *peleterianum* (Mérat) Fiori (**F**)
Hieracium peleterianum Mérat (**BT**) (**HLH**)

Family: Compositae

Colour: yellow
Distribution: western and central Alps, western Europe to Lapland, Baltic
Frequency: uncommon

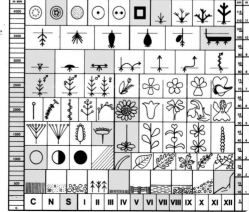

Hieracium murorum L., 1753
Hieracium murorum L. (**F**) (**BT**) (**HLH**)
Hieracium silvaticum Zahn (*)

Family: Compositae

Colour: yellow
Distribution: central and southern Europe, Iceland, western Asia, Siberia
Frequency: very common

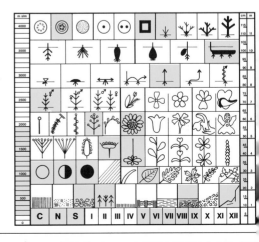

Gagea fistulosa (Ram. ex DC.), Ker-Gawler, 1816
Gagea fistulosa Ker-Gawl. (**F**)
Gagea fistulosa (Ram.) Ker-Gawl. (**HLH**)
Gagea liottardi R. & S. (*****)

Family: Liliaceae

Colour: yellow
Distribution: Alps, Apennines, Sicily, Pyrenees, Carpathians, Balkans, Crimea
Frequency: common

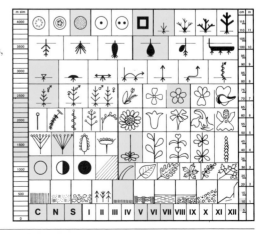

Lychnis alpina L., 1753
Lychnis alpina L. (**F**)
Viscaria alpina (L.) G. Don fil. (**BT**)
Silene liponeura Neumayr (**HLH**)

Family: Caryophyllaceae

Colour: red, purple,
pinkish red
Distribution: Alps, Tuscan
Apennines, Pyrenees,
Scotland, Scandinavia,
Iceland, Greenland, north-
eastern America, northern
Asia
Frequency: rare

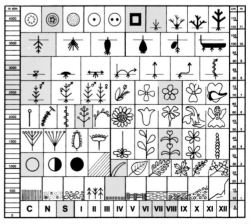

Paeonia officinalis L., 1753, subsp. *officinalis*

Paeonia officinalis L. var. *foeminea* L. (**F**)
Paeonia officinalis L. em. Gouan (**BT**)
Paeonia officinalis L. (**HLH**)

Family: Paeoniaceae

Colour: red, purple, pink
Distribution: Alps, Apennines, Asia Minor, Armenia
Frequency: rare
Properties: (*flowers and petals*) antispasmodic, sedative, narcotic; (*seeds*) emetic cathartic; to be used with prudence

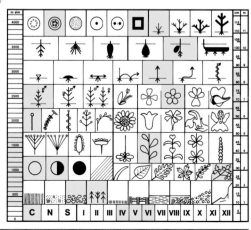

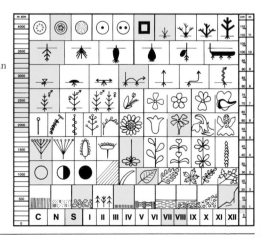

Sempervivum montanum L., 1753, subsp. ***montanum***
Sempervivum montanum L. (**F**) (**BT**) (**HLH**)

Family: Crassulaceae

Colour: dark red, violet red, violet pink, violet purple
Distribution: Alps, Tuscan Apennines, Apuan Alps, Pyrenees, Carpathians, Corsica
Frequency: uncommon

Polygonum bistorta L.

Family: Polygonaceae

Colour: pink
Distribution: upland regions throughout Europe including Britain
Frequency: locally common

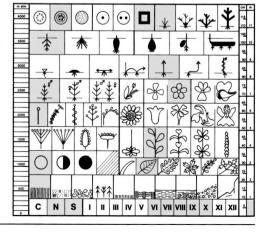

Primula minima L., 1753
Primula minima L. (**F**) (**HLH**)

Family: Primulaceae
Colour: red, purple red, pink
Distribution: eastern Alps, south-east Europe, Sudeten mtns, Carpathians, Balkans, Bulgaria
Frequency: rare

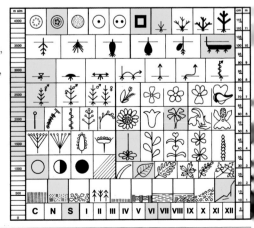

Juncus jacquinii L., 1767
Juncus jacquinii L. (**F**) (**HLH**)

Family: Juncaceae

Colour: brownish red, brownish purple, brownish black

Distribution: Alps, northern Apennines, Carpathians

Frequency: rare

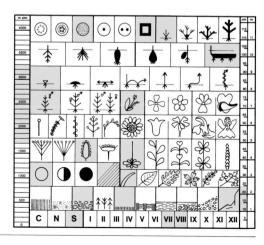

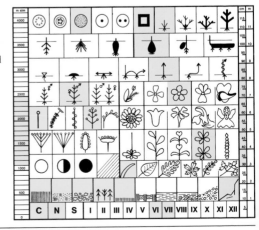

Lilium bulbiferum L., 1753 var. ***croceum*** (Chaix) Pers., 1786

Lilium bulbiferum L. var. *croceum* (Chaix) Fiori (**F**)
Lilium croceum Chaix (**HLH**)

Family: Liliaceae

Colour: red, orange, reddish orange, yellowy orange
Distribution: central and southern Europe
Frequency: rare

Cypripedium calceolus L., 1753
Cypripedium Calceolus L. (**F**) (**HLH**)

Family: Orchidaceae

Colour: (*three tepals*)
brownish red, purplish red,
ochre red; (*labellum*) yellow
Distribution: Alps,
northern Apennines,
Europe, Caucasus, Siberia,
Korea, China
Frequency: very rare
Properties: (*rhizome*)
antispasmodic, neurotonic,
anticolitic

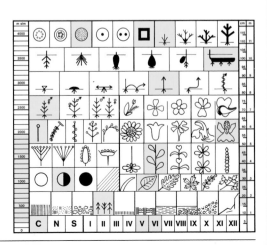

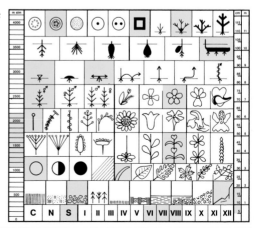

Silene vallesia L., 1759, subsp. ***vallesia***
Silene vallesia L. var. *typica* (**F**)
Silene vallesia L. (**BT**) (**HLH**)

Family: Caryophyllaceae

Colour: pink
Distribution: Alps, Apuan Alps, central Apennines
Frequency: rare

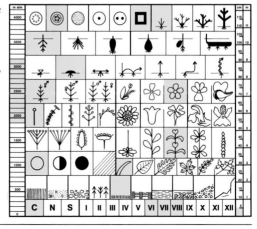

Silene acaulis (L.) Jacq., 1762
Silene acaulis L. var. *vulgaris* Reichenb. (**F**)
Silene acaulis (L.) Jacq. (**BT**) (**HLH**)

Family: Caryophyllaceae

Colour: pink, purplish red
Distribution: Alps, central and Tuscan Apennines, arctic Europe, Iceland, Pyrenees, Carpathians, Asia, Greenland, North America
Frequency: common

Saponaria ocymoides L., 1753

Saponaria ocymoides L. var. *tipica* (**F**)
Saponaria ocymoides L. (**BT**) (**HLH**)

Family: Caryophyllaceae

Colour: pink, red, purple
Distribution: southern
Europe, Asia Minor
Frequency: common
Properties: (*leaves*) *external
use:* anti-eruptive and
fluidizing, disinfectant;
poisonous in large doses

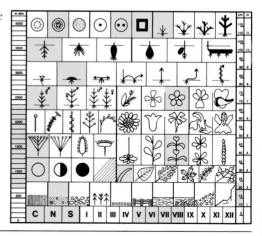

Dianthus superbus L., 1755
Dianthus superbus L. var. *typicus* (**F**)
Dianthus superbus L. (**BT**) (**HLH**)

Family: Caryophyllaceae

Colour: pink, purplish pink, violet pink, purplish violet
Distribution: Alps, northern Apennines, Europe, Siberia, central Asia
Frequency: very rare
Properties: (*petals*) sudorific, diuretic

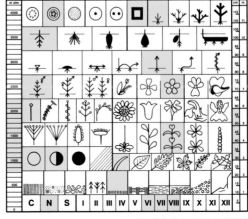

Thalictrum aquilegifolium L., 1753
Thalictrum aquilegifolium L. (**F**) (**BT**) (**HLH**)

Family: Ranunculaceae

Colour: pink, pale lilac, white
Distribution: Alps, Apennines, Europe, northern Russia, Asia (from the Altais to Japan)
Frequency: common

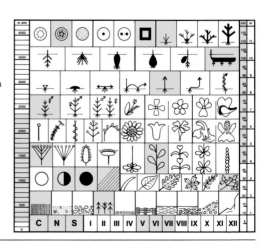

Petrocallis pyrenaica (L.) R. Br. in Aiton
Petrocallis pyrenaica R. Br. (**F**)
Petrocallis pyrenaica (L.) R. Br. (**BT**) (**HLH**)
Draba pyrenaica L. (*)

Family: Cruciferae

Colour: violet pink, lilac, white
Distribution: Alps, Pyrenees, Carpathians
Frequency: rare

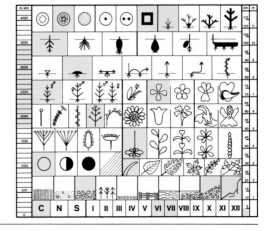

Rhodiola rosea L., 1753
Sedum roseum Scop. (**F**)
Sedum rosea (L.) Scop. (**BT**) (**HLH**)
Sedum rhodiola DC. (*****)

Family: Crassulaceae

Colour: pink, yellowish
pink, yellow
Distribution: Alps,
northern Apennines,
northern Europe; in Britain,
confined mainly to northern
England and Scotland
Frequency: locally
common

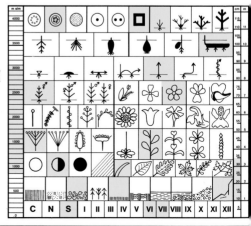

Rosa vosagiaca Desportes, 1828

Rosa glauca Vill. var. *typica* (**F**)
Rosa vosagiaca Desp. (**BT**) (**HLH**)
Rosa afzeliana Fries subsp. *vosagiaca* (Desp.) R. Keller & Gams (*)
Rosa glauca Pourret subsp. *reuteri* (Godet) Hayek (*)

Family: Rosaceae

Colour: bright pink, pink
Distribution: Alps, Apennines, Europe (not Iceland), north-east Africa, Asia, North America
Frequency: uncommon
Properties: (*leaves*) astringent; (*petals*) laxative, refreshing; (*fruits*) astringent, corrective, vitaminic; used for jams and liqueurs

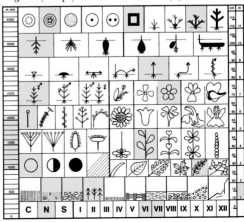

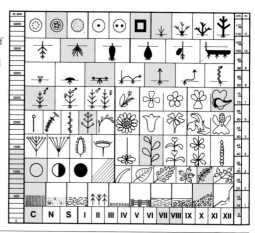

Astragalus alpinus L.

Family: Leguminosae

Colour: pale pink
Distribution: mountains of
southern and central
Europe; in Britain, confined
to a few Scottish mountains
Frequency: locally
common but rare in Britain

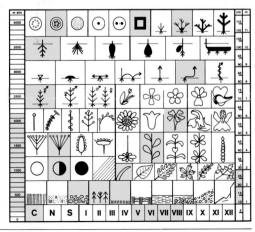

Daphne cneorum L., 1753
Daphne cneorum L. var. *typica* (**F**)
Daphne cneorum L. (**BT**) (**HLH**)

Family: Thymelaeaceae

Colour: pink, bright pink, red

Distribution: Alps, northern and central Apennines, central Europe, Bulgaria, Ukraine

Frequency: uncommon

Properties: (*fruit*) drastic purgative

Erica herbacea L., 1753
Erica carnea L. (**F**) (**BT**) (**HLH**)

Family: Ericaceae

Colour: pink, violet pink, red, dark red, purple

Distribution: Alps, high Apennines, central and southern Europe, Balkans, Carpathians

Frequency: common

Properties: (*leaves, flowering tips*) sudorific, diuretic, galattogogue

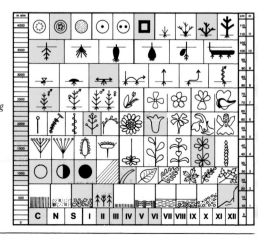

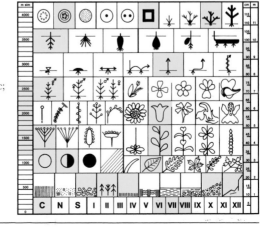

Rhododendron hirsutum L., 1753
Rhododendron hirsutum L. (**F**) (**BT**) (**HLH**)

Family: Ericaceae

Colour: pink, deep pink
Distribution: Alps, Slovenia
Frequency: uncommon (endemic)
Properties: *(leaves)* antirheumatic, antilithiasic; to be used with caution

Loiseleuria procumbens (L.) Desv., 1813

Azalea procumbens L. (**F**)
Loiseleuria procumbens (L.) Desv. (**BT**) (**HLH**)

Family: Ericaceae

Colour: pink, red, pale pink
Distribution: Alps,
Pyrenees, Britain, Lapland,
Iceland, Carpathians, Asia
(from the Altais to Japan),
arctic America, Greenland
Frequency: common
Properties: (*leaves*)
diaphoretic, diuretic; to be
used with caution

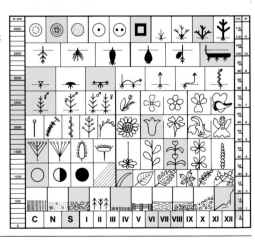

Primula farinosa L., 1753, subsp. ***farinosa***
Primula farinosa L. var. *typica* (**F**)
Primula farinosa L. (**BT**) (**HLH**)

Family: Primulaceae

Colour: pink, lilac pink, pale purple, dark purple, reddish lilac
Distribution: Alps, mountainous and alpine regions of Europe, Asia and the Americas; in Britain, confined to northern England
Frequency: locally common

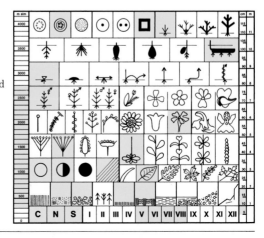

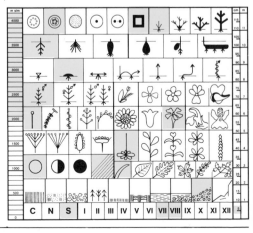

Androsace alpina (L.) Lam., 1778
Androsace alpina Lam. var. *typica* (**F**)
Androsace alpina (L.) Lam. (**BT**) (**HLH**)
Androsace glacialis Hoppe (*****)

Family: Primulaceae

Colour: pink, pale pink, white, deep pink, reddish pink, red
Distribution: Alps
Frequency: uncommon

Armeria alpina (DC.) Willd.
Armeria maritima subsp. *alpina* (Willd.) P. Silva
Armeria vulgaris W. var. *alpina* (W.) Fiori (**F**)
Armeria alpina (DC.) Willd. (**BT**) (**HLH**)
Satice montana Miller (*****)

Family: Plumbaginaceae

Colour: pink, purple, red, lilac pink
Distribution: Alps, Pyrenees, Carpathians, Balkans
Frequency: rare (endemic)
Properties: (*leaves and aerial parts*) astringent

Note: the familiar coastal plant thrift *Armeria maritima* also grows in mountain areas in many parts, including northern Britain

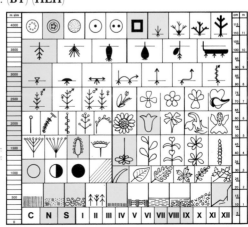

Veronica urticifolia Jacq., 1773
Veronica urticifolia Jacq. (**F**)
Veronica latifolia L. em. Scop. (**BT**)
Veronica latifolia L. (**HLH**)

Family: Scrophulariaceae

Colour: pink, lilac, white, blue

Distribution: Alps, Apuan Alps, Apennines, Pyrenees, Balkans, Carpathians, Urals

Frequency: common

Properties: *(flowering tips)* tonic stomachic

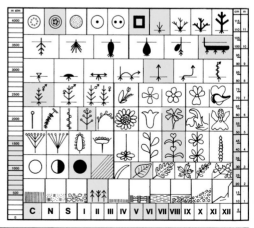

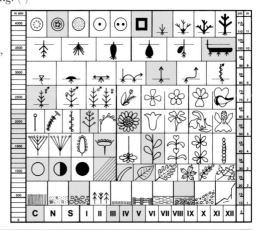

Petasites paradoxus (Retz.) Baumg., 1816
Petasites paradoxus Baumg. var. *typicus* (**F**)
Petasites paradoxus (Retz.) Baumg. (**BT**) (**HLH**)
Petasites niveus (Vill.) Baumg. (*****)

Family: Compositae

Colour: pink, pale pink
Distribution: Alps, Pyrenees, Jura, Carpathians, Balkans
Frequency: rare

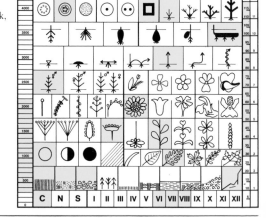

Centaurea scabiosa L., 1753
Centaurea scabiosa L. var. *calcarea* (Jord.) Fiori (**F**)
Centaurea scabiosa L. (**BT**) (**HLH**)

Family: Compositae

Colour: pink, reddish pink, purple
Distribution: Valle d'Aosta, Maritime Alps, central Alps, France
Frequency: uncommon
Properties: (*roots*) bitter stomachic, carminative, antithermic

Bulbocodium vernum L., 1753
Bulbocodium vernum L. (**F**)
Colchicum bulbocodium Ker-Gawl. (**HLH**)

Family: Liliaceae

Colour: pink, lilac pink, lilac

Distribution: Piedmontese and Maritime Alps, Pyrenees, Serbia, southern Russia, Caucasus

Frequency: rare

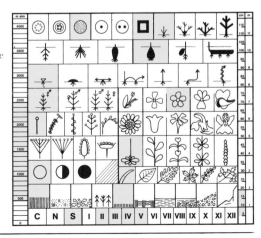

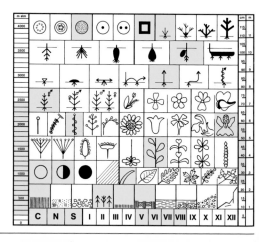

Gymnadenia conopsea (L.) R. Br. in Aiton, 1813
Gymnadenia conopsea R. Br. var. *typica* (**F**)
Gymnadenia conopsea (L.) R. Br. (**HLH**)

Family: Orchidaceae

Colour: pink, purplish
pink, violet pink, reddish
violet, lilac, violet
Distribution: Apennines,
Britain, Europe, Caucasus,
Asia Minor, northern Iran,
Asia, Siberia, northern
China, Korea, Japan,
Manchuria
Frequency: common
Properties: (*plant*) anti-
epileptic; (*tubers*) emollient,
antispasmodic

Hepatica nobilis Miller, 1768
Anemona hepatica L. var. *macrantha* Goir. (**F**)
Hepatica nobilis Schreber (**BT**)
Hepatica trioloba Gilib. (**HLH**)

Family: Ranunculaceae

Colour: violet, blue, pink, white
Distribution: Europe, Siberia, Japan, North America
Frequency: common

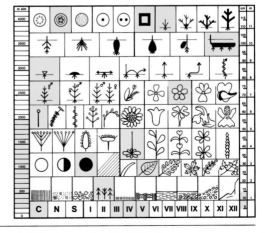

Pulsatilla montana (Hoppe) Reichenb., 1832
Anemone pulsatilla L. var. *montana* (Hpe.) Fiori (**F**)
Pulsatilla montana (Hoppe) Reichenb. (**BT**) (**HLH**)
Anemone montana Hoppe (*****)

Family: Ranunculaceae

Colour: dark violet, dark bluish violet
Distribution: Alps, Carpathians, Balkans
Frequency: rare
Properties: (*whole plant*) anti-arthritic, sudorific, febrifuge, narcotic, sedative; to be used with caution

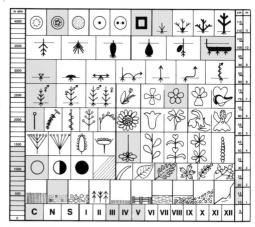

Bartsia alpina L.

Family: Scrophulariaceae

Colour: violet, bluish violet
Distribution: mountains and arctic regions throughout Europe; in Britain, confined to northern England and parts of Scotland
Frequency: locally common

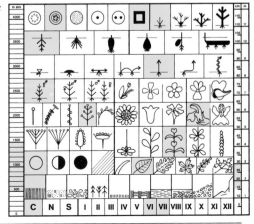

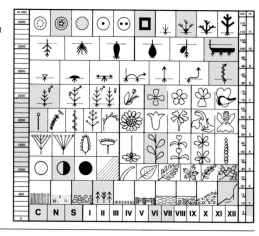

Clematis alpina (L.) Miller, 1768, subsp. ***alpina***

Clematis alpina Mill. (**F**)
Clematis alpina (L.) Miller (**BT**) (**HLH**)
Atragene alpina L. (*)

Family: Ranunculaceae

Colour: violet, bluish violet
Distribution: Alps and
Emilian Apennines, central
and southern Europe,
Carpathians, Balkans
Frequency: uncommon
Properties: (*leaves*)
revulsive, irritant

Oxytropis halleri Bunge ex Koch, 1843, subsp. ***halleri***

Astragalus uralensis L. var. *sericeus* (Lam.) Fiori (**F**)
Oxytropis halleri Bunge (**BT**) (**HLH**)
Oxytropis sericea (DC.) Simonkai, not Nutt. (*)
Astragalus sericeus Lam.
pro parte (*)

Family: Leguminosae

Colour: violet, lilac violet, bluish violet, reddish violet, lilac
Distribution: Alps, Scotland, Pyrenees, Carpathians
Frequency: rare

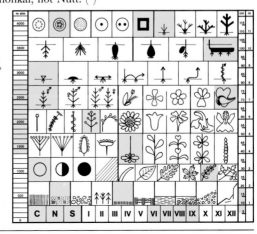

Viola lutea Hudson
Viola hirta L. var. *beraudii* (Bor.) Fiori (**F**)
Viola wolfiana W. Becker (**BT**)
Viola suavis M. Bieg (**HLH**)
Viola austriaca A. & J.
Kerner (*)
Viola beraudii Boreau (*)
Viola sepincola Jordan (*)
Viola pontica W. Becker (*)

Family: Violaceae

Colour: bluish violet,
sometimes with yellow
Distribution: uplands and
mountains of central and
western Europe; in Britain,
occurs mainly in northern
England and Scotland
Frequency: locally
common

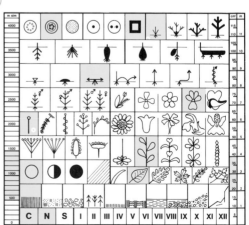

Viola calcarata L., 1753
Viola calcarata L. var. *typica* (**F**)
Viola calcarata L. (**BT**) (**HLH**)

Family: Violaceae

Colour: violet, purple, pink
(also yellow and very
occasionally white)
Distribution: Alps,
Balkans
Frequency: common
Properties: (*root*) irritant,
vomitory; (*flowers*) bechic,
pectoral, diaphoretic,
emollient, laxative (light)

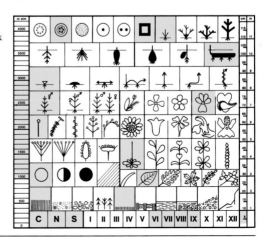

Gentianella campestris (L.) Börner, 1912

Gentiana campestris L. (**F**) (**BT**) (**HLH**)
Gentianella campestris (L.) Börner (*)

Family: Gentianaceae

Colour: violet, bluish violet, lilac violet (white)
Distribution: Alps, Apennines and much of central and northern Europe; in Britain, occurs mostly in the north
Frequency: locally common
Properties: (*plant*) bitter, febrifuge; it provides a substance used as a dye or an aromatic

Note: the corolla has four lobes

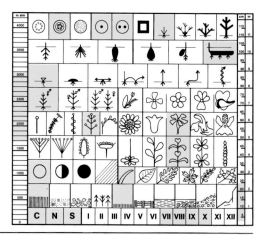

Gentianella germanica (Willd.) E. F. Warburg in Clapham,
Tutin & E. F. Warburg, 1952
Gentiana amarella L. var. *germanica* (W.) Fiori (**F**)
Gentiana germanica Willd.
(**BT**) (**HLH**)

Family: Gentianaceae

Colour: violet, reddish
violet
Distribution: Alps, central
and western Europe; in
Britain, very local in the
Chilterns
Frequency: local
Properties: (*plant*)
aperitive, stomachic; it
provides an aromatic and
colouring substance

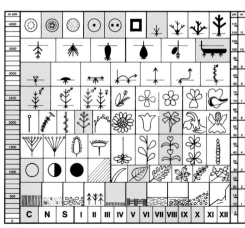

Linaria alpina (L.) Miller, 1768
Linaria alpina Mill. (**F**)
Linaria alpina (L.) Miller (**BT**) (**HLH**)

Family: Scrophulariaceae

Colour: violet, bluish violet
Distribution: Alps, central
Apennines, Pyrenees,
Carpathians, Balkans
Frequency: uncommon

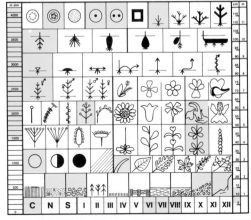

Campanula glomerata L., 1753
Campanula glomerata L. (**F**) (**BT**) (**HLH**)

Family: Campanulaceae

Colour: violet, bluish violet, azure violet, purplish azure, azure
Distribution: widespread in central Europe and northwards to southern England
Frequency: locally common

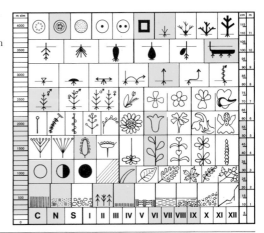

Campanula spicata L., 1753
Campanula spicata L. (**F**) (**BT**) (**HLH**)

Family: Campanulaceae

Colour: violet, bluish
violet, azure violet, azure
Distribution: Alps, Apuan
Alps, northern and central
Apennines
Frequency: rare

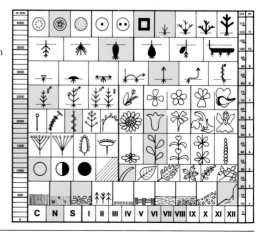

Campanula rhomboidalis L., 1753

Campanula rhomboidalis L. var. *typica* (**F**)
Campanula rhomboidalis L. (**BT**) (**HLH**)

Family: Campanulaceae

Colour: violet, azure violet, azure, blue, bluish violet
Distribution: western and central Alps, Pyrenees
Frequency: common

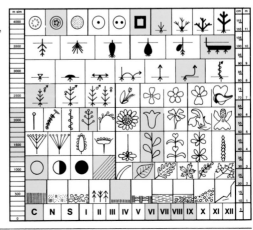

Campanula cochlearifolia Lam., 1785

Campanula rotundifolia L. var. *cochlearifolia* (Lam.) Fiori (**F**)
Campanula cochlearifolia Lam. (**BT**) (**HLH**)
Campanula pusilla Haenke (*****)

Family: Campanulaceae

Colour: violet, blue, bluish lilac, azure, lilac, white
Distribution: Alps, northern and central Apennines, Pyrenees, Carpathians, Balkans
Frequency: common

Dark rampion   105

Phyteuma ovatum Honckeny, 1782
Phyteuma halleri All. var. *typicum* (**F**)
Phyteuma ovatum Honckeny (**BT**) (**HLH**)
Phyteuma halleri All. (*****)

Family: Campanulaceae

Colour: violet, blackish violet, bluish black
Distribution: Alps, northern Apennines, Pyrenees, Balkans
Frequency: common

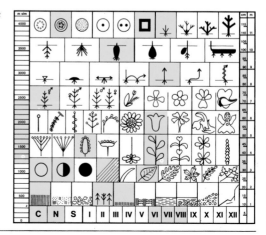

Aster alpinus L., 1753
Aster alpinus L. var. *typicus* (**F**)
Aster alpinus L. (**BT**) (**HLH**)

Family: Compositae

Colour: violet, bluish violet, lilac, azure violet, pink, white
Distribution: Alps, northern and central Apennines, Europe, Carpathians, Balkans, Urals, northern Russia, Caucasus, Siberia, Armenia, Iran, central Asia, Kamchatka, North America
Frequency: common

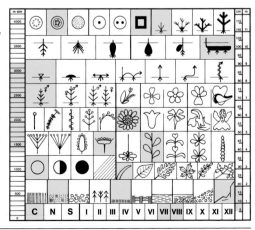

Homogyne alpina (L.) Cass

Family: Compositae

Colour: bluish purple
Distribution: mountains of
central and southern
Europe; in Britain, confined
to a few sites in Scotland
where probably introduced
Frequency: locally
common

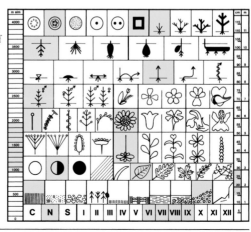

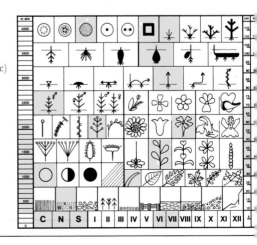

Iris latifolia (Miller) Voss in Siebert et Voss, 1895
Iris xiphioides Ehrh. (*)

Family: Iridaceae

Colour: bright blue
Distribution: Pyrenees,
Spanish Cordillera
Frequency: rare (endemic)

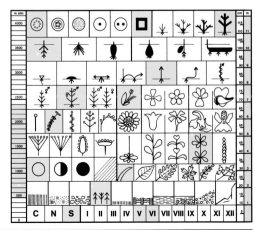

Larix decidua Miller, 1768
Larix decidua Miller (**F**) (**BT**) (**HLH**)

Family: Pinaceae
Colour: purplish red, violet red
Distribution: Alps, Carpathians
Frequency: very common
Properties: (*resin*) anticatarrhal, diuretic, haemostatic

Note : a tree growing to a height of 52m (170ft); it is the only European conifer to shed its leaves in winter; the female flowers are red, the male ones brown; used for its wood and the extraction of oil of turpentine

Dianthus glacialis Haenke in Jacq., 1789
Dianthus glacialis Haenke (**F**) (**BT**) (**HLH**)

Family: Caryophyllaceae

Colour: purple, pink, bright pink
Distribution: eastern Alps, Carpathians
Frequency: rare (endemic)

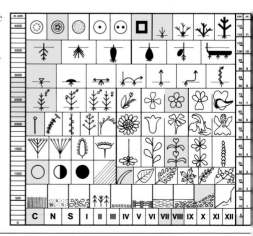

Corydalis solida (L.) Swartz, 1817
Corydalis solida Sw. var. *typica* (**F**)
Corydalis solida (Miller) Clairv. (**BT**)
Corydalis solida (L.) Swartz (**HLH**)

Family: Papaveraceae

Colour: purple, dark red, pink, white
Distribution: Alps, Apennines and much of Europe, naturalised in parts of Britain
Frequency: local
Properties: (*tubercles*) sedative, narcotic, diaphoretic, depurative; to be used with caution

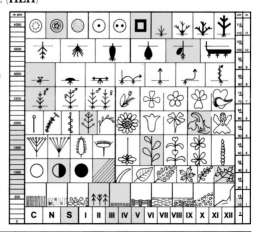

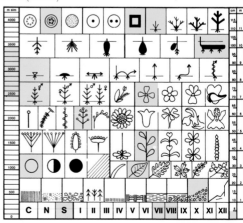

Thlaspi rotundifolium (L.) Gaudin, 1829, subsp. ***rotundifolium***

Thlaspi rotundifolium Gaud. var. *corymbosum* Gaud. (**F**)
Thlaspi rotundifolium (L.) Gaudin ssp. *corymbosum* (J. Gay) Gremli (**BT**)
Thlaspi corymbosum (Gay) Rchb. (**HLH**)

Family: Cruciferae

Colour: dark lilac, pale violet, purple, bright pink
Distribution: Val d'Aosta, Monte Rosa, Mt Cenis, Savoy, Swiss Alps
Frequency: rare

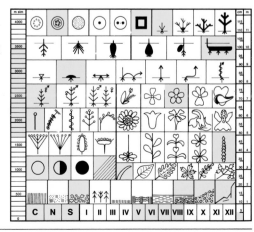

Saxifraga oppositifolia L., 1753
Saxifraga oppositifolia L. var. *distans* Ser. (**F**)
Saxifraga oppositifolia L. (**BT**) (**HLH**)

Family: Saxifragaceae

Colour: purple, lilac red, pink
Distribution: western and Maritime Alps, Sierra Nevada, Pyrenees, Carpathians, northern Europe, northern Britain, northern Siberia, North America, Greenland
Frequency: uncommon ·

Saxifraga retusa Gouan, 1773
subsp. ***augustana*** (Vacc.) D. A. Webb, 1963

Saxifraga retusa Gouan var. *augustana* L. Vacc. (**F**)
Saxifraga retusa Gouan (**BT**)
Saxifraga purpurea All.
(**HLH**)

Family: Saxifragaceae

Colour: purple red, purple
Distribution: western and
Maritime Alps, Savoy
Frequency: rare

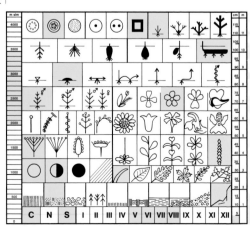

Astragalus danicus Retz., 1783
Astragalus hypoglottis L. var. *danicus* (Retz.) Fiori (**F**)
Astragalus danicus Retz. (**HLH**)
Astragalus hypoglottis auct.? not L. (*****)

Family: Leguminosae

Colour: purple, bluish
purple, bluish violet, violet
Distribution: Maritime
and Piedmontese Alps,
Ireland, central Europe,
Britain, Caucasus, Siberia,
central Asia, North America
Frequency: rare

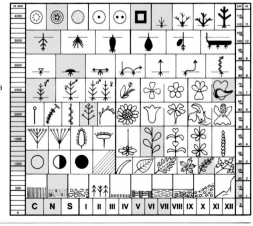

Astragalus monspessulanus L., 1753
Astragalus monspessulanus L. var. *typicus* (**F**)
Astragalus monspessulanus L. (**BT**) (**HLH**)

Family: Leguminosae

Colour: bright purple, reddish purple, violet purple
Distribution: southern Europe to Sicily, Balkans, Rumania, north-west Africa (Tunisia, Algeria)
Frequency: common

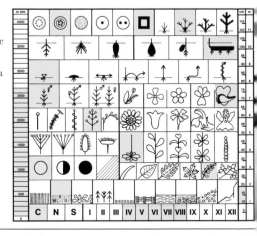

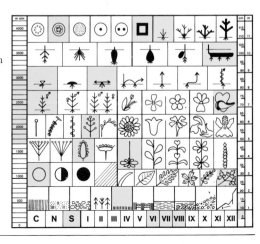

Trifolium alpinum L., 1753
Trifolium alpinum L. (**F**) (**BT**) (**HLH**)

Family: Leguminosae

Colour: purple, reddish purple, pink

Distribution: Alps, Apuan Alps, Ligurian Apennines, Tuscan and central Apennines, Pyrenees

Frequency: common

Epilobium fleischeri Hochst., 1826
Epilobium dodonaci Vill. var. *alpinum* Burn (**F**)
Epilobium fleischeri Hochst. (**BT**) (**HLH**)

Family: Onagraceae

Colour: purple, bright pink
Distribution: Alps
Frequency: common

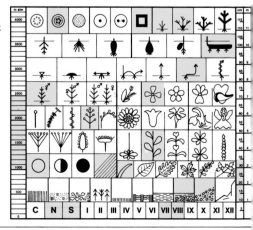

Soldanella alpina L., 1753
Soldanella alpina L. var. *typica* (**F**)
Soldanella alpina L. (**BT**) (**HLH**)

Family: Primulaceae

Colour: lilac, azure violet, bluish violet, blue, violet
Distribution: Alps, Apennines, Pyrenees
Frequency: common
Properties: (*roots*) purgative

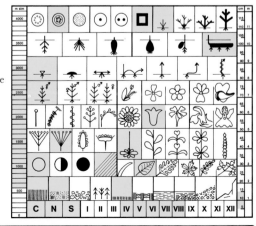

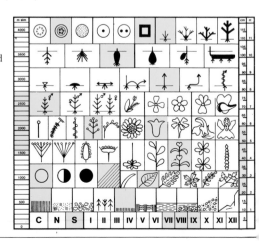

Gentiana purpurea L., 1753

Gentiana purpurea L. var. *typica* (**F**)
Gentiana purpurea L. (**BT**) (**HLH**)

Family: Gentianaceae

Colour: purple, reddish purple

Distribution: western and central Alps, northern Apennines, Apuan Alps, Scandinavia, Carpathians, Caucasus, Kamchatka

Frequency: rare

Properties: (*roots*) antifermentative, aperitive, bitter tonic, digestive, febrifuge, vermifuge; its tincture is used to treat digestive troubles

Scutellaria alpina L., 1753, subsp. *alpina*

Family: Labiatae

Colour: purple, azure violet, bluish violet, violet
Distribution: Alps, northern and central Apennines, Pyrenees, Balkans, central Asia, Siberia, Altais, Urals, Turkestan, Ukraine
Frequency: rare

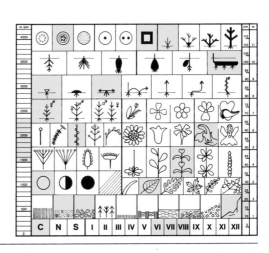

Pedicularis verticillata L., 1753
Pedicularis verticillata L. (**F**) (**BT**) (**HLH**)

Family:
Scrophulariaceae

Colour: purple, reddish
purple, violet purple, pink
Distribution: Alps, Apuan
Alps, northern and central
Apennines, Europe,
northern and central Asia,
North America
Frequency: common

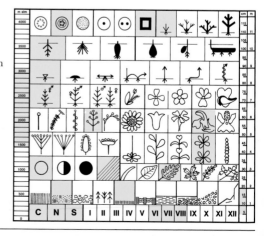

Carduus defloratus L., 1759
Carduus defloratus L. (**F**) (**BT**) (**HLH**)

Family: Compositae

Colour: purple, red, pink
Distribution: Alps, Apennines, central Europe, Carpathians
Frequency: common

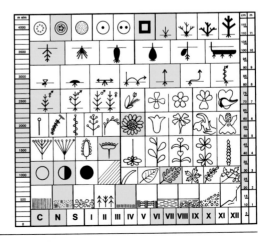

Cirsium eriophorum (L.) Scop., 1772

Cirsium eriophorum Scop. (**F**)
Cirsium eriophorum (L.) Scop. (**BT**) (**HLH**)

Family: Compositae

Colour: purple, lilac, reddish purple
Distribution: western Alps, Apennines, western Europe northwards to southern England
Frequency: locally common

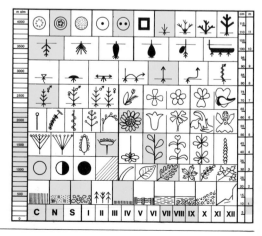

Centaurea uniflora Turra, 1765
subsp. **nervosa** (Willd.) Bonnier & Layens, 1894
Centaurea uniflora Turra var. *nervosa* (W.) Fiori (**F**)
Centaurea nervosa Willd. (**BT**) (**HLH**)
Centaurea uniflora Turra
ssp. *nervosa* Rouy (*)

Family: Compositae

Colour: purple, violet,
reddish purple
Distribution: Alps,
northern Apennines,
Balkans, Carpathians
Frequency: rare

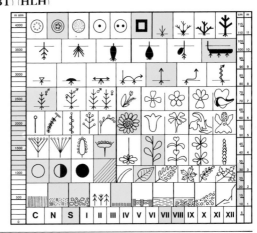

Allium schoenoprasum L., 1753
Allium schoenoprasum L. var. *sibiricum* (L.) Fiori (**F**)
Allium schoenoprasum L. (**HLH**)

Family: Liliaceae

Colour: lilac, pink, pinkish purple, violet purple, pinkish violet, bluish pink, azure violet

Distribution: Alps, central and southern Europe, Siberia, Himalayas, Japan, North America

Frequency: common

Properties: (*bulb*) stimulant, digestive, carminative, antiseptic, powerful intestinal microbicide, balsamic, diuretic, antinicotinic, vermifuge

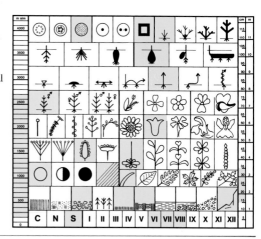

Lilium martagon L., 1753
Lilium martagon L. var. *typicum* (**F**)
Lilium martagon L. (**HLH**)

Family: Liliaceae

Colour: purple, pink, purplish pink, purplish red, white
Distribution: Alps, Apennines, central and southern Europe, Caucasus, Asia to Japan
Frequency: common
Properties: (*bulb*) diuretic, antiarthritic; *external use:* emollient, resolutive

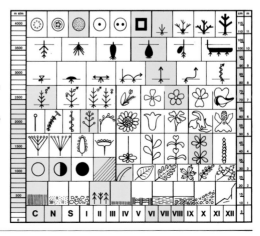

Orchis mascula (L.) L., 1755
Orchis mascula L. (**F**) (**HLH**)

Family: Orchidaceae

Colour: purple, purplish
red, pink, red, lilac
Distribution: Sardinia,
central and southern
Europe, central Russia, Asia
Minor, Caucasus, Iran,
Urals, northern Africa
(Morocco, Algeria)
Frequency: common
Properties: (*tuber*)
emollient, excipient

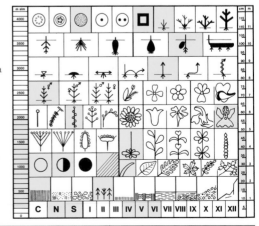

Orchis incarnata L., 1755

Orchis latifolia L. var. *incarnata* (L.) Fiori (**F**)
Orchis incarnata L. (**HLH**)
Orchis latifolia L. sec. Pugsl. (*)
Dactylorhiza incarnata (L.) Soó, 1962

Family: Orchidaceae

Colour: purple, pink, carmine red, violet pink
Distribution: central and southern Europe, Siberia, central Asia, Mongolia
Frequency: uncommon
Properties: (*tuber*) emollient, excipient

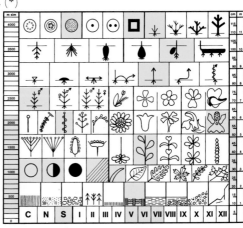

Nigritella nigra (L.) Reichenb. fil., 1851
Nigritella nigra Rchb. (**F**)
Nigritella nigra (L.) Rchb. (**HLH**)
Nigritella angustifolia Rich. (*****)

Family: Orchidaceae

Colour: purple, purplish black, reddish black, reddish brown, red, pink
Distribution: Alps, Apennines, Pyrenees, Jura, Carpathians, Balkans, Scandinavia
Frequency: locally fairly common
Properties: (*aerial parts*) aphrodisiac

Note: vanilla-scented flowers

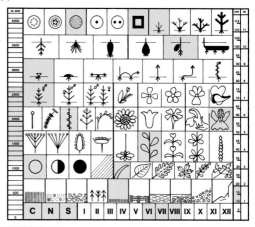

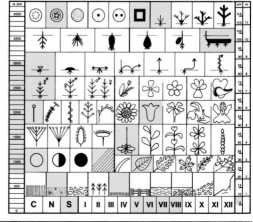

Gentiana acaulis L., 1753
Gentiana acaulis L. var. *latifolia* Gr. & Godr. (**F**)
Gentiana kochiana Perr. & Song. (**BT**) (**HLH**)
Gentiana excisa C. Presl. (*)

Family: Gentianaceae
Colour: blue, dark azure, dark blue
Distribution: Alps, Apuan Alps, northern and central Apennines, Pyrenees, Balkans, Carpathians
Frequency: common
Properties: (*whole plant*) bitter aperitive, tonic, stomachic, febrifuge

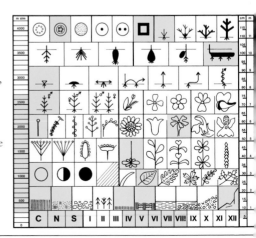

Gentiana verna L., 1753
Gentiana verna L. (**F**) (**BT**) (**HLH**)

Family: Gentianaceae

Colour: dark azure, violet, blue, dark blue
Distribution: Alps, Apennines, Ireland, England, Pyrenees, Balkans, Carpathians, Asia Minor, Caucasus, Turkestan, Afghanistan, Altais, eastern Siberia, Mongolia
Frequency: common
Properties: bitter aperitive

Gentiana brachyphylla Vill., 1779
Gentiana verna L. var. *brachyphylla* (Vill.) Fiori (**F**)
Gentiana brachyphylla Vill. (**BT**) (**HLH**)

Family: Gentianaceae

Colour: azure, dark azure, pale blue, blue
Distribution: Alps, Sierra Nevada, Pyrenees
Frequency: rare
Properties: bitter aperitive

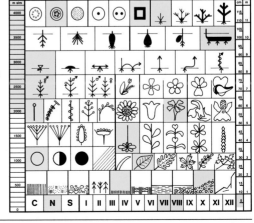

Gentiana bavarica L., 1753
Gentiana bavarica L. var. *typica* (**F**)
Gentiana bavarica L. (**BT**) (**HLH**)

Family: Gentianaceae

Colour: azure, dark blue, violet
Distribution: Alps
Frequency: uncommon (endemic)
Properties: bitter aperitive

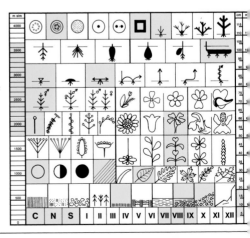

Echium vulgare L., 1753

Echium vulgare L. var. *typicum* (**F**)
Echium vulgare L. (**BT**) (**HLH**)

Family: Boraginaceae·

Colour: blue, violet blue, violet, pink
Distribution: Europe, western Asia, western Siberia, northern Africa, North America (naturalised)
Frequency: very common
Properties: *(flowering spike)* sudorific, pectoral, anti-inflammatory, anticatarrhal

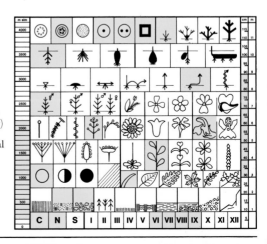

Veronica alpina L.
Veronica chamaedrys L. var. *typica* (**F**)
Veronica chamaedrys L. (**BT**) (**HLH**)

Family: Scrophulariaceae

Colour: blue
Distribution: mountains throughout Europe; in Britain, confined to the Scottish highlands
Frequency: locally common

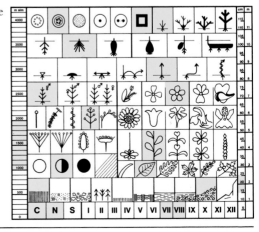

Phyteuma hemisphericum L., 1753

Phyteuma hemisphericum L. var. *graminifolium* (Sieb.) Fiori (**F**)
Phyteuma hemisphericum L. (**BT**) (**HLH**)

Family: Campanulaceae

Colour: blue, bluish violet
Distribution: Alps, Apennines, Pyrenees
Frequency: common

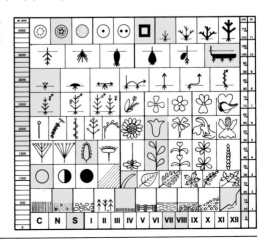

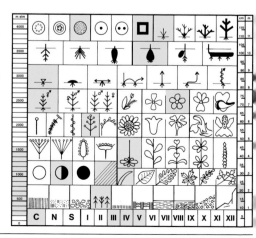

Scilla bifolia L., 1753
Scilla bifolia L. var. *typica* (**F**)
Scilla bifolia L. (**HLH**)

Family: Liliaceae

Colour: blue, azure, bluish
azure, purplish blue, white,
bluish pink, pink
Distribution: central and
southern Europe, Sicily and
Sardinia, Asia Minor,
Caucasus
Frequency: uncommon
Properties: (*bulb*)
cardiotonic, diuretic,
expectorant

Aquilegia alpina L., 1753
Aquilegia alpina L. var. *typica* (**F**)
Aquilegia alpina L. (**BT**) (**HLH**)

Family: Ranunculaceae

Colour: azure, violet azure, blue
Distribution: Piedmontese Alps, Tuscan Apennines
Frequency: rare

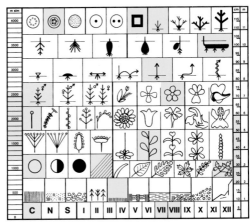

Linum perenne L., 1753, subsp. ***alpinum*** (Jacq.) Ockendon, 1967

Linum perenne L. var. *alpinum* (Jacq.) Fiori (**F**)
Linum alpinum Jacq. (**BT**) (**HLH**)

Family: Geraniaceae

Colour: azure
Distribution: Alps, Apennines, southern and central Europe, northern England
Frequency: locally common
Properties: (*whole plant*) equine purgative

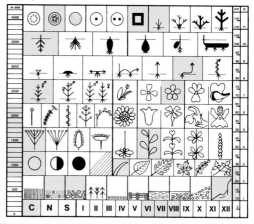

Gentiana clusii Perr. et Song 1855

Gentiana acaulis L. var. *vulgaris* Neilr. (**F**)
Gentiana clusii Perr. & Song (**BT**) (**HLH**)

Family: Gentianaceae

Colour: azure, intense azure
Distribution: Alps, Apennines
Frequency: rare
Properties: (*whole plant with leaves*) bitter aperitive, tonic, stomachic, febrifuge

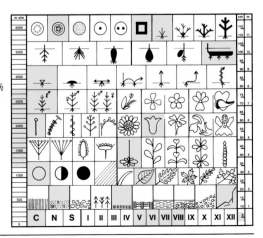

Gentianella ciliata (L.) Borkh. 1796
Gentiana ciliata L. var. *typica* (**F**)
Gentiana ciliata L. (**BT**) (**HLH**)

Family: Gentianaceae
Colour: azure, pale blue, blue
Distribution: Alps, northern and central Apennines, Pyrenees, central England (very local), Balkans, Carpathians, Ukraine, Caucasus
Frequency: rare

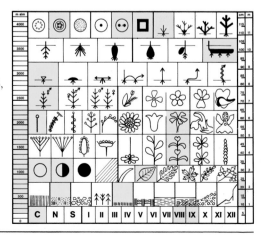

Myosotis alpestris F. W. Schmidt, 1794

Myosotis alpestris Schm. var. *typica* (**F**)
Myosotis alpestris F. W. Schmidt (**BT**) (**HLH**)
Myosotis pyrenaica Auct. (*****)
Myosotis sylvatica Hoffm.
subsp. *alpestris* (F. W.
Schmidt) Gams (*****)

Family: Boraginaceae

Colour: azure, dark azure,
blue

Distribution: Alps,
Apennines, Pyrenees,
Balkans; in Britain, very
local in northern England
and Scottish highlands

Frequency: common

Properties: (*aerial parts*)
slightly astringent

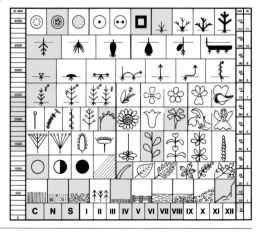

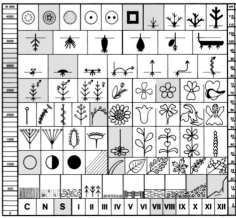

Eritrichium nanum (L.) Schrader ex Gaudin, 1828, subsp. ***nanu***

Eritrichium nanum Schrad. (**F**)
Eritrichium nanum (L.) Gaud. (**BT**)
Eritrichium nanum (Amann) Schrader (**HLH**)
Myosotis nana Amann (*)

Family: Boraginaceae

Colour: azure
Distribution: Alps,
Carpathians
Frequency: rare

Ajuga pyramidalis L., 1753
Ajuga genevensis L. var. *pyramidalis* (L.) Fiori (**F**)
Ajuga pyramidalis L. (**BT**) (**HLH**)

Family: Labiatae

Colour: azure, violet azure, violet blue, pink
Distribution: Alps, Ligurian Apennines, Europe; in Britain, very local in northern England and Scotland
Frequency: common

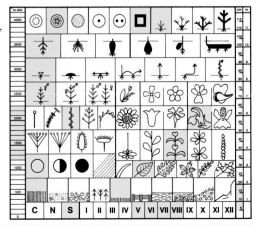

Veronica fruticans Jacq., 1762
Veronica fruticulosa L. *var. fruticans* (Jacq.) Fiori (**F**)
Veronica fruticans Jacq. (**BT**) (**HLH**)
Veronica saxatilis Scop. (*****)

Family: Scrophulariaceae

Colour: azure, blue
Distribution: Alps, northern and central Apennines, northern Europe, Carpathians, Balkans; in Britain, very local in Scottish mountains
Frequency: common

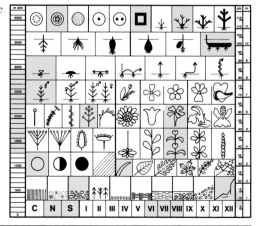

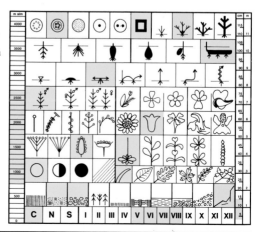

Globularia cordifolia L., 1753
Globularia cordifolia L. (**F**) (**BT**) (**HLH**)

Family: Globulariaceae
Colour: azure, azure grey,
azure pink, blue, white
Distribution: Alps, Apuan
Alps, central and southern
pennines, Balkans,
Carpathians, Crimea
Frequency: common

Note: single flowers, irregularly
bellshaped

Globularia nudicaulis L., 1753
Globularia nudicaulis L. (**F**) (**BT**) (**HLH**)

Family: Globulariaceae

Colour: azure, blue
Distribution: Alps, northern Apennines, Pyrenees
Frequency: common
Properties: (*leaves*) diuretic, bitter tonic, stomachic, light purgative, diaphoretic

Note: single flower; irregular bellshaped corolla

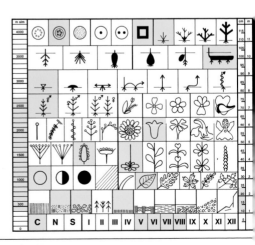

Campanula cenisia L., 1763
Campanula cenisisa L. (**F**) (**BT**) (**HLH**)

Family: Campanulaceae

Colour: azure, blue, azure lilac
Distribution: Valtellina, Piedmont, Ticinese Alps (endemic)
Frequency: very rare

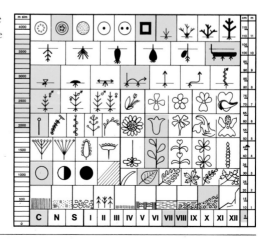

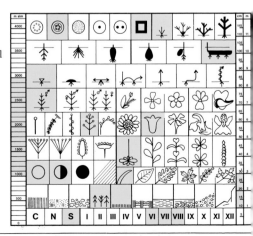

Campanula barbata L., 1759
Campanula barbata L. var. *typica* (**F**)
Campanula barbata L. (**BT**) (**HLH**)

Family: Campanulaceae

Colour: azure, pale azure, bluish lilac, blue, white
Distribution: Alps, central and southern Europe, Norway, Sudeten mtns, western Carpathians
Frequency: common

Centaurea triumfetti All., 1773

Centaurea montana L. var. *triumfetti* (All.) Fiori (**F**)
Centaurea triumfetti All. (**BT**)
Centaurea triumfettii All. (**HLH**)
Centaurea axillaris Willd. (*)
Centaurea lugdunensis
Gaudin (*)

Family: Compositae

Colour: azure, blue, violet,
purplish red
Distribution: Alps,
Apennines, Pyrenees,
Carpathians, Balkans, Asia
Minor, Caucasus, northern
Africa (Morocco)
Frequency: common

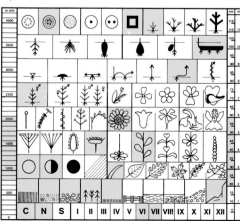

Listera cordata (L.) R. Br.

Family: Orchidaceae

Colour: greenish brown
Distribution: widespread in Europe, mostly in upland regions
Frequency: locally common

Note: Cryptogamous (flowerless) plant, carrying the spores on its strobili

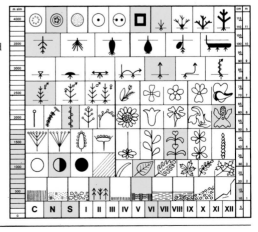

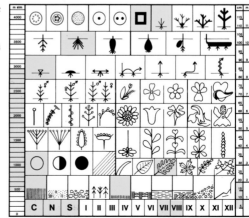

Botrychium lunaria (L.) Schwartz in Schrader, 1802

Botrychium lunaria Sw. var. *typicum* (**F**)
Botrychium lunaria (L.) Sw. (**BT**) (**HLH**)

Family: Ophioglossaceae

Colour: green
Distribution: Europe,
northern Asia, Japan, North
America, Patagonia, Chile,
Australia
Frequency: uncommon
Properties: (*whole plant*)
cicatrizing, used to treat
sluggish wounds and
cancerous sores

Note: Cryptogamous (flowerless)
plant carrying its spores on the bunch
of spikes

Pinus uncinata Miller ex Mirbel in Buffon, 1806
Pinus mugo Turra var. *uncinata* (Ram.) Fiori (**F**)
Pinus mugo Turra ssp. *uncinata* Domin (**HLH**)

Family: Pinaceae

Colour: green, yellowish green
Distribution: western Alps, Piedmontese Apennines, Pyrenees
Frequency: rare
Properties: (*young branches*) balsamic, anticatarrhal, antiputrefacient

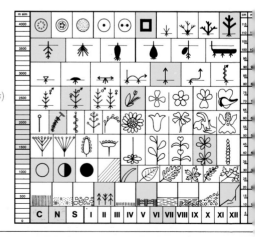

Juniperus communis L., 1753, subsp. *communis*

Juniperus communis L. var. *typica* (**F**)
Juniperus communis L. (**BT**) (**HLH**)

Family: Cupressaceae

Colour: green, yellowish green

Distribution: Europe, Algeria, northern and western Asia to the Himalayas, North America

Frequency: common

Properties: (*leaves*) diuretic, anti-eczematous; (*fruits*) aromatic, tonic digestive, sudorific, antigout, antirheumatic, diuretic, disinfectant of the urinary ducts; used in the production of gin

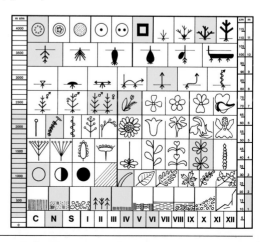

Salix herbacea L., 1753
Salix herbacea L. (**F**) (**BT**) (**HLH**)

Family: Salicaceae

Colour: green, yellowish green

Distribution: Alps, central Apennines, Pyrenees, Carpathians, Balkans, Scotland, Greenland, North America

Frequency: common

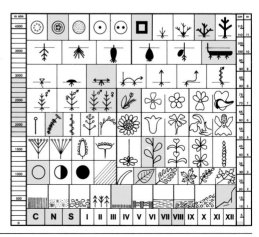

Oxyria digyna (L.) Hill, 1768
Oxyria digyna Hill (**F**)
Oxyria digyna (L.) Hill (**BT**) (**HLH**)

Family: Polygonaceae

Colour: grassy green
Distribution: Alps, arctic
Europe, Spain, Corsica,
Bulgaria
Frequency: common

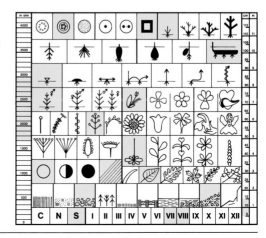

Vaccinium myrtillus L.

Family: Ericaceae

Colour: green
Distribution: widespread
throughout Europe
Frequency: locally
abundant

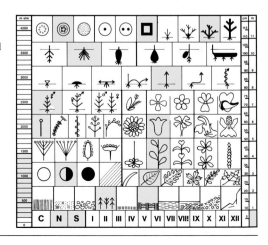

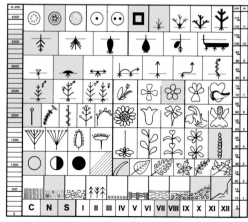

Minuartia sedoides (L.) Hiern, 1899

Alsine sedoides Kitt. (**F**)
Minuartia sedoides (L.) Hiern (**BT**) (**HLH**)
Cherleria sedoides L. (*)

Family: Caryophyllaceae

Colour: greenish yellow
Distribution: Alps,
Pyrenees, Carpathians,
Scotland
Frequency: common

Orthilia secunda (L.) House

Family: Pyrolaceae

Colour: green
Distribution: upland regions throughout most of Europe
Frequency: locally common

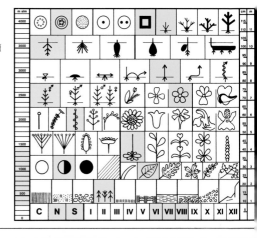

Alchemilla alpina L.

Family: Rosaceae

Colour: green
Distribution: northern Europe; in Britain, found mainly in Scotland
Frequency: locally common
Properties: (*leaves*) astringent, diuretic, soothing, sedative; *external use* haemostatic detergent

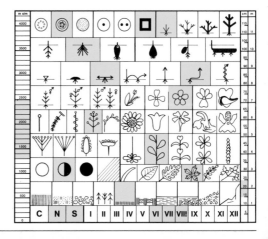

Arctostaphylos alpinus (L.) Sprengel

Family: Ericaceae

Colour: green
Distribution: northern
Europe including northern
Scotland
Frequency: local

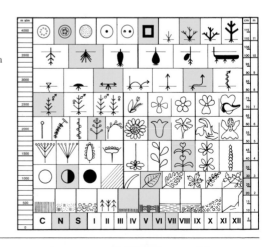

Rhamnus pumilus Turra, 1764

Rhamnus pumila Turra var. *typica* (**F**)
Rhamnus pumila Turra (**BT**) (**HLH**)

Family: Rhamnaceae

Colour: green, yellowish green

Distribution: Alps, central and southern Apennines, Sardinia, southern Europe (from Spain to Albania), north-west Africa (Atlas)

Frequency: uncommon

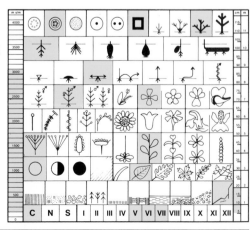

Sambucus racemosa L., 1753
Sambucus racemosa L. (**F**) (**BT**) (**HLH**)

Family: Caprifoliaceae

Colour: green, yellowish green
Distribution: Alps, northern Apennines, central Europe, Baltic, Carpathians, Bulgaria
Frequency: very common
Properties: (*bark*) purgative; (*flowers*) sudorific, diuretic, emollient, resolutive; (*fruits*) laxative; berries are used in jam-making

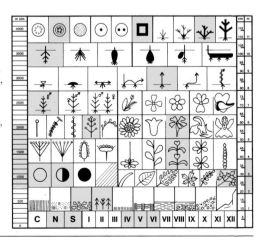

Artemisia genipi Weber in Stechm., 1775
Artemisia genipi Weber var. *typica* (**F**)
Artemisia genipi Weber (**BT**) (**HLH**)
Artemisia spicata Wulfen (*****)

Family: Compositae

Colour: green, silvery grey, yellow, brown
Distribution: Alps, Pyrenees
Frequency: rare
Properties: (*flowering parts*) bitter, stimulant, stomachic, digestive, fortifying, excitant, tonic, sudorific, febrifuge, balsamic expectorant, antiscorbutic; used in liqueurs and sweets

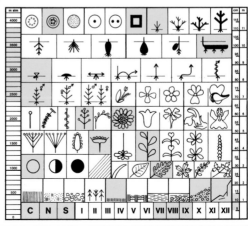

Phleum phleoides (L.) Karsten, 1881
Phleum phleoides Simonk., 1886, var. *typicum* (**F**)
Phleum boehmeri Wibel (**HLH**)

Family: Graminaceae

Colour: green
Distribution: Alps,
Apennines, Mt Etna,
Europe, northern, western
and central Asia, northern
Africa (Algeria)
Frequency: uncommon

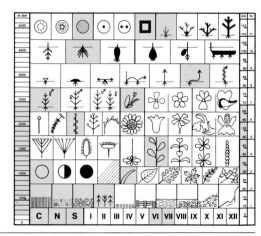

Veratrum album L., 1753
Veratrum album L., var. *lobelianum* (Bernh.) Fiori (**F**)
Veratrum lobelianum Bernh. (**HLH**)

Family: Liliaceae

Colour: green, pale green, yellowish green
Distribution: Alps, Apennines, Pyrenees, Carpathians, central Asia, Siberia, Kamchatka, Altais, Japan, Arctic
Frequency: common
Properties: (*rhizome*) purgative and vomitory (poisonings), antithermic, depressive of the bulbar centres; (*whole plant*) antidiarrhoeic. *Not for family use.* It is a poisonous plant which is sometimes mistaken for *Gentiana lutea* which is not poisonous

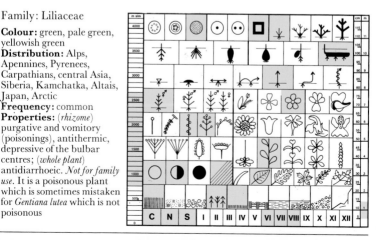

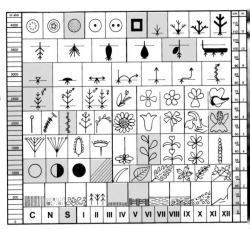

Coeloglossum viride (L.) Hartman, 1820
Coeloglossum viride Hartm. (**F**)
Coeloglossum viride (L.) Hartman (**HLH**)

Family: Orchidaceae
Colour: green, yellowish green
Distribution: Alps, Apennines, Europe, Russia, Urals, Siberia, Crimea, Caucasus, Turkestan, eastern Asia, North America
Frequency: locally common

GLOSSARY OF
PHARMACEUTICAL TERMINOLOGY

Antidiarrhoetic: controls and moderates intestinal movements.

Anti-eruptive: controls and eliminates the exodus through the mouth of gases which form in the stomach during digestion.

Antihaemorrhagic: controls and stops haemorrhages and blood loss.

Antihelmintic: see Vermifuge.

Antipyretic: see Febrifuge.

Antiscropholous: treats and prevents the swelling and suppuration of lymphatic glands.

Antiseptic: a remedy against putrefaction and germ infections.

Antispasmodic: a sedative of the nervous system: controls muscular contractions.

Antithermic: see Febrifuge.

Aperitive: stimulates the appetite.

Aphrodisiac: a substance which increases sexual stimulus and response.

Aromatic: slightly excitant.

Astringent: contracts tissues, making them more taut, and controls mucous secretions.

Balsamic: controls and reduces strong secretions by soothing the inflamed mucous tissues of the respiratory and urinary ducts.

Bechic: a remedy for coughs and bronchial afflictions.

Bitter: a substance which aids digestion.

Cardiotonic: increases the heart's energy.

Cardiovascular: aids the workings of the circulatory system.

Carminative: diminishes and expels abnormal gases from the digestive system and soothes pain.

Corrective: a substance used to correct the taste and smell of drugs.

Corroborant: a tonic for the whole organism.

Depurative: encourages the expulsion (through sweat or urine) of harmful substances and impurities thus purifying both blood and organism.

Detergent: a substance used to clean wounds and ulcers and encourage cicatrization.

Diaphoretic (sudorific): causes and increases transpiration.

Diuretic: encourages the secretion of urine.

Emetic: provokes vomiting.

Emmenagogue: re-establishes and regulates the menses.

Emollient: soothes and eliminates inflammation of tissues, thus protecting and softening mucous membranes.

Excipient: inert liquid or dough-like substance used to add body to drugs.

Excitant: stimulant.

Expectorant: facilitates the expulsion of bronchial secretions.

Febrifuge: reduces body temperature.

Fluidizer: helps dissolve food in the stomach.

Galactagogue: aids and increases milk secretion.

Haemosolvent: liquefies the blood.

Haemostatic: coagulant, halts haemorrhages.

Hypertensive: lowers the blood pressure.

Insecticide: kills insects.

Irritant: produces inflammation.

Laxative: mild stimulant of intestinal evacuation.

Narcotic: soothing substance which induces sleep.

Neurotonic: tonic for the nervous system.

Pectoral: a drug which treats the tissues affected by ailments of the respiratory ducts.

Purgative: a strong stimulant of intestinal evacuations.

Refreshing: relieves thirst and lowers the temperature of the organism; it also aids the intestinal movements.

Resolutive: a substance which can dispel obstructions, granulomas, inflammations and most ailments in general.

Revulsive: a drug to be used externally; it causes an increase in the blood afflux to the area it is applied to, followed by artificial irritation, which in turn determines the decrease of irritation of the affected internal organ.

Rubefacient: a drug to be used externally; it induces and increases the flow of blood to the surface, causing reddening of the skin.

Sedative: soothes internal and external inflammations.

Stimulant: excites the functions of the nervous, muscular and circulatory systems.

Stomachic: it improves and reinforces the functioning of the stomach.

Sudorific: activates and increases transpiration.

Sympathicolithic: annuls the effect of the sympathetic nervous system.

Tincture: alcoholic infusion of vegetal substances.
Tonic: increases the vitality and energy of the organism.
Topic: a drug applied only and directly to the exterior of the afflicted area.

Vermifuge: destroys and eliminates intestinal worms.
Vomitory: see Emetic.

FURTHER READING

Aichelle, D. and R. and Schwegler, H. and A., *Hamlyn Guide to Wild Flowers of Britain and Europe*, Hamlyn (1992)

Baumann, Hellmut, *Greek Wild Flowers – and Plant Lore in Ancient Greece*, Herbert Press (1993)

Beckett, E., *Wildflowers of Majorca, Minorca and Ibiza*, Balkema (1988)

Blamey, Marjorie and Grey-Wilson, Christopher, *The Illustrated Flora of Britain and Northern Europe*, Hodder (1989)

Bonner, A., *Plants of the Balearic Islands* (1992)

Chinery, *Field Guide to the Plant Life of Britain and Europe*, Kingfisher (1987)

Clapham, A. R., Tutin, T. G. and Moore, D. M., *Flora of the British Isles*, CUP (1989)

Clapham, A. R., Tutin, T. G. and Warburg, E. F., *Excursion Flora of the British Isles*, CUP (1981)

Council of Europe, *Vegetation of the Alps* (1983)

Crozier, J. and Matschke, A., *Flowers of Andorra* (1992)

Davies, Paul and Gibbons, Bob, *Field Guide to the Wild Flowers of Southern Europe*, Crowood (1993)

Ellis, G., *Flowering Plants of Wales*, National Museum of Wales (1983)

Elvers, I., *Our Flora in Colour/ Var Flora i Farg*, Norstedts (1991)

Faegri, K. and Iversen, J., *Which Plant? – Identification Keys for the Northwest European Pollen Flora*, Wiley

Feilberg, J., Fredskild, B. and Holt, S., *Flowers of Greenland – Gronlands Blomster*, Arctic Book/Map Svs (1984)

Fitter, *Wild Flowers of Britain and Northern Europe – Collins New Generation Guide*, OOP (1987)

Fitter, Richard, Fitter, Alastair and Blamey, Marjorie, *Collins Pocket Guide: The Wild Flowers of Britain and Northern Europe*, Harper Collins (1985)

Forey, P., *The Pocket Guide to Wild Flowers of the British Isles and Northern Europe*, Dragon's World (1992); *Wild Flowers of the British Isles and Northern Europe*, Dragon's World (1991)

Gibbons, Bob and Brough, Peter, *The Hamlyn Photographic Guide to the Wild Flowers of Britain and Northern Europe*, Hamlyn (1992)

Gjaerevoll, O., *Maps of Distribution of Norwegian Vascular Plants, Vol 1 Alpine Plants*, Tapir (1990)

Grey-Wilson, Christopher and Blamey, Marjorie, *The Alpine Flowers of Britain and Northern Europe*, Harper Collins (1986)

Hackney, P., *Flora of North-East of Ireland*, Irish Studies Inst. (1992)

Hayward, John, *A New Key to Wild Flowers*, CUP (1987)

Innes, C., *Wild Flowers of Spain, Vols 1–3*, Cockatrice (1987)

Jalas, Jaako and Suominen, Julia, *Atlas Flora Europaeae – Distribution of Vascular Plants in Europe, Vols 1–3*, Paris NHM (1988)

Keble Martin, *The New Concise British Flora* (1991)

Kent, D. H., *List of Vascular Plants of the British Isles*, BSBI (1992)

Kristinsson, H., *A Guide to the Flowering Plants and Ferns of Iceland*, Arctic Book/Map Svs (1987)

Landolt, Elias, *Our Alpine Flora – Swiss Alpine Club*, SAC (1991)

Lipert and Podleich, *Collins Nature Guides: Wild Flowers*, Harper Collins (1994)

Meikle, R. D., *The Flora of Cyprus, Vols 1 and 2*, Kew

Moore, *The Mitchell Beazley Pocket Guide to Wild Flowers*, Mitchell Beazley (1992)

Nylen, B., *Nordic Flora/Nordens Flora*, Norstedts (1992)

Parish/Parish, *Wild Flowers – A Photographic Guide*, Dovecote (1989)

Perring, *RSNC Guide to British Wild Flowers*, Hamlyn (1989)

Perring and Farrell, *British Red Data Book 1: Vascular Plants*, RSNC (1983)

Perring, F. H. and Walters, S. M., *Atlas of the British Flora*, BSBI (1990); *The Macmillan Field Guide to British Wildflowers*, Macmillan (1989)

Phillips, R., *Wild Flowers of Britain*, Pan (1977)

Polunin, Oleg, *A Concise Guide to the Flowers of Britain and Europe*, OUP (1987); *Collins Photoguide to Wild Flowers of Britain and Northern Europe*, Harper Collins (1987)

Polunin, Oleg and Huxley, Anthony, *Flowers of the Mediterranean*, Chatto & Windus (1990)

Polunin, Oleg and Smythies, B. E., *Flowers of South-west Europe – A Field Guide*, OUP (1988)

Polunin, Oleg and Stainton, Adam, *Flowers of the Himalaya*, OUP (1988)

Press, Bob, *Photographic Field Guide: Wild Flowers of Britain and Europe*, New Holland (1993)

Rose, Francis, *The Wild Flower Key: British Isles – NW Europe – with Keys to Plants not in Flower*, Warne (1991)

Ross, R. I., *Pocket Guide to Irish Wild Flowers*, Appletree Press (1987)

Schauer and Caspari, *A Field Guide to the Wild Flowers of Britain and Europe*, Harper Collins (1982)

Sfikas, G., *The Wild Flowers of Crete*, Efstathiadis (1987); *Wild Flowers of Cyprus*, Efstathiadis (1993); *Wild Flowers of Greece*, Efstathiadis (1979)

Stace, Clive A., *New Flora of the British Isles*, CUP (1987)

Turland, N. J., Chilton, L. and Press, J. R., *Flora of the Cretan Area – Annotated Checklist and Atlas*, (1993)

Tutin, T. G., Heywood, V. H., Burges, N. H. et al, *Flora Europaea, Vols 1–5*, CUP (1968–93)

Uniyal, M. R., *Medicinal Flora of Garwhal Himalayas*, (1989)

Webb, D. A., *An Irish Flora*, Dundalgan Press (1977)

Webb, D. and Gornall, R., *Saxifrages of Europe*, Helm (1989)

ALPHABETICAL INDEX OF LATIN NAMES

Numbers given refer to colour plates.
Synonyms are in italics

ALPHABETICAL INDEX OF
VERNACULAR NAMES

Also published by David & Charles

WILD HERBS OF BRITAIN AND EUROPE
Jacques de Sloover & Martine Goossens

This *Naturetrek* guide to herbs with culinary, medicinal and aromatic uses is an ideal identification handbook and fascinating home herbal. It shows in superb colour photographs 144 species of herb as they will typically be found in the wild, for realistic comparison, and groups them simply by the colour of their flowers, with the page edges correspondingly coloured for quick selection.

For each species a coloured chart with easily recognizable symbols compacts a wealth of botanical detail to enable the scientific confirmation of the initial visual identification. This key summarizes the herb's properties, which parts are efficacious and when they should be gathered, where it grows and when it flowers. Each herb is identified by its botanical Latin name, family and most usual common name.

The at-a-glance field guide is supplemented by a straightforward introduction to the use of herbs, and several informative appendices. These include botanical notes on peculiar features not indicated by the pictograms, and alternative common names, a guide to the methods of preparing herbal treatments, lists of herbs and their properties, ailments and their herbal remedies, and guidance on the use of herbs in the kitchen. A glossary, list of further reading and full index make this a self-contained reference, or an ideal companion in further study.

MUSHROOMS & TOADSTOOLS OF BRITAIN AND EUROPE
V. Norris

Anyone interested in collecting mushrooms, whether to study them scientifically or simply to enrich an everyday diet with their many delicious flavours and pure nutritional value, will find this *Naturetrek* guide an invaluable companion. Grouping of the species by predominant colour, shown on the page edges, instantly gives a selection of possible identifications for a specimen. The superb colour photographs of the living fungi, in their typical habitats and seen under natural lighting, provide the most realistic comparison possible, whilst the unique pictogram provides detailed botanical information at a glance to verify the identification scientifically and provide a clear statement of whether the species is edible, inedible or poisonous.

Richard Mabey, in his foreword, explains how to go about your first mushroom hunt, and tells of the many facets of this rewarding hobby. The introduction describes the main genera, their life-cycle and distribution, provides guides to the cultivation of fungi, to mycology (or the study of fungi), explains how to distinguish edible from inedible or poisonous varieties, and describes the nutritional value of mushrooms. A further reading list, etymology of scientific terms and various indexes complete this as *the* field guide for all mushroom collectors.

MINERALS & GEMSTONES OF THE WORLD
G. Brocardo

This handy *Naturetrek* identification guide offers a superb collection of photographs and compacts a wealth of information in a unique way that is useful to both experienced collectors and beginners alike. The 156 specimens are shown in individual colour photographs, and ordered by their overall colour which is shown on the page corners for quick reference. Pictographic tables use easily identified symbols to allow instant selection from a mass of detailed information on the characteristics and properties which can be used to identify and classify each mineral – including its geometrical structure, and reaction to physical and chemical tests. The key to the symbols is printed on a bookmark for easy reference alongside each page.

Before the main field guide there is a useful introduction on how to recognize and collect minerals, their origin and formation, structure and properties, their classification, and how to prepare them for preservation. There is also a full explanation of the tests which may be used to help identify specimens using only a few simple pieces of equipment, and a summary of more advanced laboratory techniques. A glossary, bibliography and indexes complete the work to make it both a practical handbook and a condensed reference manual.

MOUNTAIN FLOWERS OF BRITAIN AND EUROPE
S. Stefenelli

This *Naturetrek* guide to the mountain flowers of Britain and Europe will make a walk in the hills an exciting botanical adventure.

Thousands of enchanting flowers grow on the mountain slopes of Europe, and this book will prove an informative and useful guide for those wishing to discover more about them, whilst aiding an appreciation of their beauty and an understanding of the need for their conservation.

Recognition is made easy by comparison with 168 colour photographs of the flowers as they will typically be found. To identify a species, simply match the colour of the flower to the corresponding colour section and your task becomes easy. Once it has been clearly identified, the pictograms which accompany the plates will enable both the beginner and the serious botanist to discover at a glance the other interesting facts about the flower. The symbols used in the pictograms are quickly learned, but the clever use of flaps on the cover enable the keys to be taken next to the page for reference.

A section on habitat, a glossary of pharmaceutical terminology, a further reading list and two indexes make this a handy home reference and companion for further study as well as being the ultimate field guide.